Invertebrate Zoology

Invertebrate Zoology

Editor: Jean Wilkinson

R CALLISTO REFERENCE

www.callistoreference.com

Callisto Reference,
118-35 Queens Blvd., Suite 400,
Forest Hills, NY 11375, USA

Visit us on the World Wide Web at:
www.callistoreference.com

ISBN: 978-1-64116-017-9 (Hardback)

Cataloging-in-Publication Data

Invertebrate zoology / edited by Jean Wilkinson.
    p. cm.
Includes bibliographical references and index.
ISBN 978-1-64116-017-9
1. Invertebrates. 2. Animals. 3. Zoology. I. Wilkinson, Jean.
QL362 .I58 2018
592--dc23

Table of Contents

Permissions

Index

Preface

As a sub-part of zoology, invertebrate zoology refers to the study of animals without backbones, commonly called invertebrates. This is a vast category of animals which includes tunicates, sponges, arthropods, worms, echinoderms, molluscs, etc. There are many sub-divisions of this field namely malacology, arthropodology, helminthology, invertebrate paleontology, etc. This book presents the complex subject of invertebrate zoology in the most comprehensible and easy to understand language. It unfolds the innovative aspects of this field which will be crucial for the holistic understanding of the subject matter. For someone with an interest and eye for detail, the text covers the most significant topics in the field of invertebrate zoology.

A foreword of all Chapters of the book is provided below:

Chapter 1 - The subject concerned with the study of invertebrate animals, or animals without a vertebral column, is known as invertebrate zoology. The group of invertebrate animals is very diverse. Some of the animals classified as invertebrates are sponges, worms and arthropods. This chapter will provide an integrated understanding of invertebrate zoology; **Chapter 2 -** Invertebrates are animals that do not have a backbone or a spine. Many invertebrates possess an asymmetric body shape. The major categories of invertebrates are dealt with great details in the chapter; **Chapter 3 -** The category of invertebrates found in marine habitats is known as marine invertebrates. Echinoderm, cnidaria and tunicate are some of the examples of marine invertebrates. The aspects elucidated in this chapter are of vital importance, and provide a better understanding of marine invertebrates; **Chapter 4 -** Phylum is one of the categories in the hierarchy of biological classifications. Some of the types of phyla of invertebrates are flatworms, annelids, ctenophora and mollusca. The topics elaborated in this section will help in gaining a better perspective about the various phyla of invertebrates.

I would like to thank the entire editorial team who made sincere efforts for this book and my family who supported me in my efforts of working on this book. I take this opportunity to thank all those who have been a guiding force throughout my life.

Editor

An Introduction to Invertebrate Zoology

The subject concerned with the study of invertebrate animals, or animals without a vertebral column, is known as invertebrate zoology. The group of invertebrate animals is very diverse. Some of the animals classified as invertebrates are sponges, worms and arthropods. This chapter will provide an integrated understanding of invertebrate zoology.

Invertebrate Zoology

Invertebrate zoology is the subsystem of zoology that consists of the study of invertebrates, animals without a backbone (a structure which is found only in fish, amphibians, reptiles, birds and mammals.)

Invertebrates are a vast and very diverse group of animals that includes sponges, echinoderms, tunicates, numerous different phyla of worms, molluscs, arthropods and many additional phyla. Single-celled organisms or protists are usually not included within the same group as invertebrates.

Subdivisions

Invertebrates represent 97% of all named animal species, and because of that fact, this subdivision of zoology has many further subdivisions, including but not limited to:

- Arthropodology - the study of arthropods, which includes
- Arachnology - the study of spiders and other arachnids
- Entomology - the study of insects
- Carcinology - the study of crustaceans
- Myriapodology - the study of centipedes, millipedes, and other myriapods
- Cnidariology - the study of Cnidaria
- Helminthology - the study of parasitic worms.
- Malacology - the study of mollusks, which includes
- Conchology - the study of Mollusk shells.
- Limacology - the study of slugs.
- Teuthology - the study of cephalopods.
- Invertebrate paleontology - the study of fossil invertebrates

These divisions are sometimes further divided into more specific specialties. For example, within arachnology, acarology is the study of mites and ticks; within entomology, lepidoptery is the study of butterflies and moths, myrmecology is the study of ants and so on.

History

Early Modern Era

In the period starting in the late 15th century, invertebrate zoology saw growth in the number of publications made and improvement in the experimental practices associated with the field.

One of the major works to be published in the area of zoology was Conrad Gessner's Historia animalium, which was published in numerous editions from 1551 to 1587. Though it was a work more generally addressing zoology in the large sense, it did contain information on insect life. Much of the information came from older works; Gessner restated the work of Pliny the Elder and Aristotle while mixing old knowledge of the natural history of insects with his own observations.

With the invention of the Microscope in 1599 came a new way of observing the small creatures that fall under the umbrella of invertebrate. Robert Hooke, who worked out of the Royal Society in England, conducted observation of insects—including some of their larval forms—and other invertebrates, such as ticks. His Micrographia, published in 1665, included illustrations and written descriptions of the things he saw under the microscope.

Others also worked with the microscope following its acceptance as a scientific tool. Francesco Redi, an Italian physician and naturalist, used a microscope for observation of invertebrates, but is known for his work in disproving the theory of spontaneous generation. Redi managed to prove that flies did not spontaneously arise from rotting meat. He conducted controlled experiments and detailed observation of the fly life cycle in order to do so. Redi also worked in the description and illustration of parasites for both plants and animals.

Other men were also conducting research into pests and parasites at this time. Felix Plater, a Swiss physician, worked to differentiate between two types of tape worm. He also wrote descriptions of both the worms he observed and the effects these worms had on their hosts.

Following the publication of Francis Bacon's ideas about the value of experimentation in the sciences came a shift toward true experimental efforts in the biological sciences, including invertebrate zoology. Jan Swammerdam, a Dutch microscopist, supported an effort to work for a 'modern' science over blind belief in the work of ancient philosophers. He worked—like Redi—to disprove spontaneous generation using experimental techniques. Swammerdam also made a number of advancements in the study of anatomy and physiology. In the field of entomology, he conducted a number of dissections of insects and made detailed observations of the internal structures of these specimens. Swammerdam also worked on a classification of insects based on life histories; he managed to contribute to the literature proving that an egg, larva, pupa, and adult are indeed the same individual.

18th and 19th Centuries

In the 18th century, the study of invertebrates focused on the naming of species that were relevant to economic pursuits, such as agricultural pests. Entomology was changing in big ways very quickly, as many naturalists and zoologists were working with hexapods.

Work was also being done in the realm of parasitology and the study of worms. A French physician named Nicolas Andry de Bois-Regard determined that worms were the cause of some diseases. He also declared that worms do not spontaneously form within the animal or human gut; de Bois-Regard stated that there must be some kind of 'seed' which enters the body and contains the worm in some form. Antonio Vallisneri also worked with parasitic worms, specifically members of the genera *Ascaris* and *Neoascaris*. He found that these worms came from eggs. In addition, Vallisneri worked to elucidate the reproduction of insects, specifically the sawfly.

In 1735, the first edition of Carl Linnaeus' *Systema Naturae* was published; this work included information on both insects and intestinal worms. However, the tenth edition is considered the true starting point for the modern classification scheme for living things today. Linnaeus' universal system of classification made a system based on binomial nomenclature, but included higher levels of classification than simply the genus and species names. *Systema Naturae* was an investigation into the biodiversity on Earth. However, because it was based only on very few characters, the system developed by Linnaeus was an artificial one. The book also included descriptions of the organisms named inside of it.

In 1859, Charles Darwin's On the Origin of Species was published. In this book, he described his theory of evolution by natural selection. Both the work of Darwin and his contemporary, Alfred Russel Wallace —who was also working on the theory of evolution—were informed by the careful study of insects. In addition, Darwin collected many species of invertebrate during his time aboard the HMS *Beagle*; many of the specimens collected were insects. Using these collections, he was able to study sexual dimorphism, geographic distribution of species, and mimicry; all of these concepts influenced Darwin's theory of evolution. Unfortunately, a firm popular belief in the immutability of species was a major hurdle in the acceptance of the thory.

20th Century

Classification in the twentieth century shifted toward a focus on evolutionary relationships over morphological description. The development of phylogenetics and systematics based on this study is credited to Willi Hennig, a German entomologist. In 1966, his *Phylogenetic Systematics* was published; inside, Hennig redefined the goals of systematic schemes for classifying living things. He proposed that the focus be on evolutionary relationships over similar morphological features. He also defined monophyly and included his ideas about hierarchical classification. Though Hennig did not include information on outgroup comparison, he was apparently aware of the practice, which is considered important to today's systematic research.

Interesting Invertebrates

- The Japanese spider crab (Arthropoda: *Macrocheira kaempferi*) is one of the world's most impressive arthropods. The Japanese spider crab is largest known species of crab and may live up to 100 years. With a leg span of that can reach four feet, it has the longest span of any arthropod. They are typically found in the Pacific waters near Japan on the bottom of the continental shelf.

- The lion's mane jellyfish (Cnidaria: *Cyanea capillata*) is the largest known type of jellyfish. Their tentacles can reach up to 190 feet long, and they may have a bell diameter of almost 7 feet. These animals are usually found in cold northern Arctic waters and in the Northern portions of the Atlantic and Pacific Oceans.

- The giant squid (Mollusca: *Architeuthis dux*) comes from the family Architeuthidae. These squid are both the largest known cephalopod and the largest known mollusc. They can grow to a length of about 45-50 feet long. They developed large eyes, the largest of any animal, to be able to detect small amounts of bioluminescence in the dark and deep ocean where they live.

2

Invertebrates and their Types

Invertebrates are animals that do not have a backbone or a spine. Many invertebrates possess an asymmetric body shape. The major categories of invertebrates are dealt with great details in the chapter.

Invertebrate

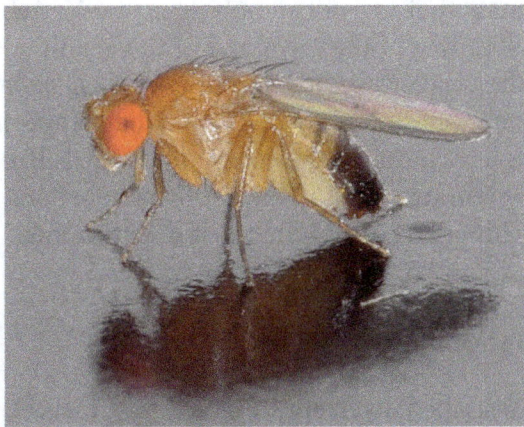

The common fruit fly, *Drosophila melanogaster*, has been used extensively for research.

Invertebrates are animals that neither possess nor develop a vertebral column (commonly known as a *backbone* or *spine*), derived from the notochord. This includes all animals apart from the sub-phylum Vertebrata. Familiar examples of invertebrates include insects; crabs, lobsters and their kin; snails, clams, octopuses and their kin; starfish, sea-urchins and their kin; jellyfish, and worms.

The majority of animal species are invertebrates; one estimate puts the figure at 97%. Many invertebrate taxa have a greater number and variety of species than the entire subphylum of Vertebrata.

Some of the so-called invertebrates, such as the Tunicata and Cephalochordata are more closely related to the vertebrates than to other invertebrates. This makes the term "invertebrate" paraphyletic and hence almost meaningless for taxonomic purposes.

Etymology

The word "invertebrate" comes from the form of the Latin word *vertebra*, which means a joint in general, and sometimes specifically a joint from the spinal column of a vertebrate. In turn the jointed aspect of *vertebra* derived from the concept of turning, expressed in the root *verto* or *vorto*, to turn. Coupled with the prefix *in-*, meaning "not" or "without".

Taxonomic Significance

The term *invertebrates* is not always precise among non-biologists since it does not accurately describe a taxon in the same way that Arthropoda, Vertebrata or Manidae do. Each of these terms describes a valid taxon, phylum, subphylum or family. "Invertebrata" is a term of convenience, not a taxon; it has very little circumscriptional significance except within the Chordata. The Vertebrata as a subphylum comprises such a small proportion of the Metazoa that to speak of the kingdom Animalia in terms of "Vertebrata" and "Invertebrata" has limited practicality. In the more formal taxonomy of Animalia other attributes that logically should precede the presence or absence of the vertebral column in constructing a cladogram, for example, the presence of a notochord. That would at least circumscribe the Chordata. However, even the notochord would be a less fundamental criterion than aspects of embryological development and symmetry or perhaps bauplan.

Despite this, the concept of *invertebrates* as a taxon of animals has persisted for over a century among the laity, and within the zoological community and in its literature it remains in use as a term of convenience for animals that are not members of the Vertebrata. The following text reflects earlier scientific understanding of the term and of those animals which have constituted it. According to this understanding, invertebrates do not possess a skeleton of bone, either internal or external. They include hugely varied body plans. Many have fluid-filled, hydrostatic skeletons, like jellyfish or worms. Others have hard exoskeletons, outer shells like those of insects and crustaceans. The most familiar invertebrates include the Protozoa, Porifera, Coelenterata, Platyhelminthes, Nematoda, Annelida, Echinodermata, Mollusca and Arthropoda. Arthropoda include insects, crustaceans and arachnids.

Number of Extant Species

By far the largest number of described invertebrate species are insects. The following table lists the number of described extant species for major invertebrate groups as estimated in the IUCN Red List of Threatened Species, *2014.*

Invertebrate group	Image	Estimated number of described species
Insects		1,000,000
Molluscs		85,000

Crustaceans		47,000
Corals		2,000
Arachnids		102,248
Velvet worms		165
Horseshoe crabs		4
Others jellyfish, echinoderms, sponges, other worms etc.		68,658
Totals:		**1,305,075**

The IUCN estimates that 66,178 extant vertebrate species have been described, which means that over 95% of the described animal species in the world are invertebrates.

Characteristics

The trait that is common to all invertebrates is the absence of a vertebral column (backbone): this creates a distinction between invertebrates and vertebrates. The distinction is one of convenience only; it is not based on any clear biologically homologous trait, any more than the common trait of having wings functionally unites insects, bats, and birds, or than not having wings unites tortoises, snails and sponges. Being animals, invertebrates are heterotrophs, and require sustenance in the form of the consumption of other organisms. With a few exceptions, such as the Porifera, invertebrates generally have bodies composed of differentiated tissues. There is also typically a digestive chamber with one or two openings to the exterior.

Morphology and Symmetry

The body plans of most multicellular organisms exhibit some form of symmetry, whether radial, bilateral, or spherical. A minority, however, exhibit no symmetry. One example of asymmetric invertebrates include all gastropod species. This is easily seen in snails and sea snails, which have helical shells. Slugs appear externally symmetrical, but their pneumostome (breathing hole) is located on the right side. Other gastropods develop external asymmetry, such as Glaucus atlanticus that develops asymmetrical cerata as they mature. The origin of gastropod asymmetry is a subject of scientific debate.

Other examples of asymmetry are found in fiddler crabs and hermit crabs. They often have one claw much larger than the other. If a male fiddler loses its large claw, it will grow another on the opposite side after moulting. Sessile animals such as sponges are asymmetrical alongside coral colonies (with the exception of the individual polyps that exhibit radial symmetry); alpheidae claws that lack pincers; and some copepods, polyopisthocotyleans, and monogeneans which parasitize by attachment or residency within the gill chamber of their fish hosts).

Nervous System

Neurons differ in invertebrates from mammalian cells. Invertebrates cells fire in response to similar stimuli as mammals, such as tissue trauma, high temperature, or changes in pH. The first invertebrate in which a neuron cell was identified was the medicinal leech, *Hirudo medicinalis*.

Learning and memory using nociceptors in the sea hare, *Aplysia* has been described. Mollusk neurons are able to detect increasing pressures and tissue trauma.

Neurons have been identified in a wide range of invertebrate species, including annelids, molluscs, nematodes and arthropods.

Respiratory System

One type of invertebrate respiratory system is the open respiratory system composed of spiracles, tracheae, and tracheoles that terrestrial arthropods have to transport metabolic gases to and from tissues. The distribution of spiracles can vary greatly among the many orders of insects, but in general each segment of the body can have only one pair of spiracles, each of which connects to an atrium and has a relatively large tracheal tube behind it. The tracheae are invaginations of the cuticular exoskeleton that branch (anastomose) throughout the body with diameters from only a few micrometres up to 0.8 mm. The smallest tubes, tracheoles, penetrate cells and serve as sites of diffusion for water, oxygen, and carbon dioxide. Gas may be conducted through the respiratory system by means of active ventilation or passive diffusion. Unlike vertebrates, insects do not generally carry oxygen in their haemolymph.

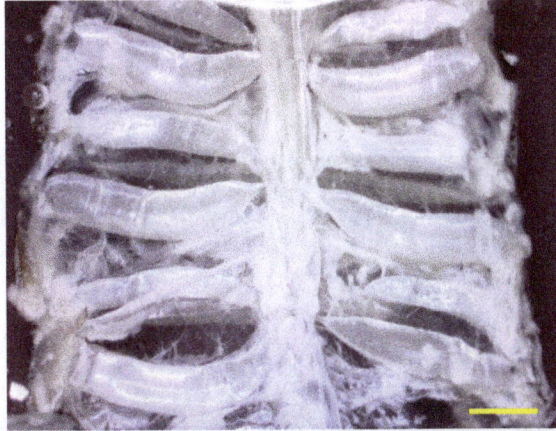

Tracheal system of dissected cockroach. The largest tracheae run across the width of the body of the cockroach and are horizontal in this image. Scale bar, 2 mm.

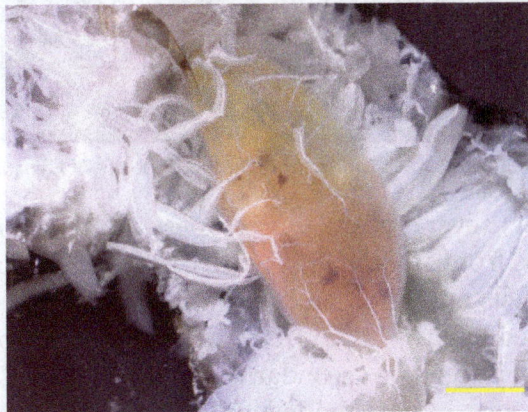

The tracheal system branches into progressively smaller tubes, here supplying the crop of the cockroach. Scale bar, 2.0 mm.

A tracheal tube may contain ridge-like circumferential rings of taenidia in various geometries such as loops or helices. In the head, thorax, or abdomen, tracheae may also be connected to air sacs. Many insects, such as grasshoppers and bees, which actively pump the air sacs in their abdomen, are able to control the flow of air through their body. In some aquatic insects, the tracheae exchange gas through the body wall directly, in the form of a gill, or function essentially as normal, via a plastron. Note that despite being internal, the tracheae of arthropods are shed during moulting (ecdysis).

Reproduction

Like vertebrates, most invertebrates reproduce at least partly through sexual reproduction. They produce specialized reproductive cells that undergo meiosis to produce smaller, motile spermatozoa or larger, non-motile ova. These fuse to form zygotes, which develop into new individuals. Others are capable of asexual reproduction, or sometimes, both methods of reproduction.

Social Interaction

Social behavior is widespread in invertebrates, including cockroaches, termites, aphids, thrips, ants, bees, Passalidae, Acari, spiders, and more. Social interaction is particularly salient in eusocial species but applies to other invertebrates as well.

Insects recognize information transmitted by other insects.

Phyla

5.0 mm

The fossil coral *Cladocora* from the Pliocene of Cyprus

The term invertebrates covers several phyla. One of these are the sponges (Porifera). They were long thought to have diverged from other animals early. They lack the complex organization found in most other phyla. Their cells are differentiated, but in most cases not organized into distinct tissues. Sponges typically feed by drawing in water through pores. Some speculate that sponges are not so primitive, but may instead be secondarily simplified. The Ctenophora and the Cnidaria, which includes sea anemones, corals, and jellyfish, are radially symmetric and have digestive chambers with a single opening, which serves as both the mouth and the anus. Both have distinct tissues, but they are not organized into organs. There are only two main germ layers, the ectoderm and endoderm, with only scattered cells between them. As such, they are sometimes called diploblastic.

The Echinodermata are radially symmetric and exclusively marine, including starfish (Asteroidea), sea urchins, (Echinoidea), brittle stars (Ophiuroidea), sea cucumbers (Holothuroidea) and feather stars (Crinoidea).

The largest animal phylum is also included within invertebrates: the Arthropoda, including insects, spiders, crabs, and their kin. All these organisms have a body divided into repeating segments, typically with paired appendages. In addition, they possess a hardened exoskeleton that is periodically shed during growth. Two smaller phyla, the Onychophora and Tardigrada, are close relatives of the arthropods and share these traits. The Nematoda or roundworms, are perhaps the second largest animal phylum, and are also invertebrates. Roundworms are typically microscopic, and occur in nearly every environment where there is water. A number are important parasites. Smaller phyla related to them are the Kinorhyncha, Priapulida, and Loricifera. These groups have a reduced coelom, called a pseudocoelom. Other invertebrates include the Nemertea or ribbon worms, and the Sipuncula.

Another phylum is Platyhelminthes, the flatworms. These were originally considered primitive, but it now appears they developed from more complex ancestors. Flatworms are acoelomates, lacking a body cavity, as are their closest relatives, the microscopic Gastrotricha. The Rotifera or rotifers, are common in aqueous environments. Invertebrates also include the Acanthocephala or spiny-headed worms, the Gnathostomulida, Micrognathozoa, and the Cycliophora.

Also included are two of the most successful animal phyla, the Mollusca and Annelida. The former, which is the second-largest animal phylum by number of described species, includes animals such as snails, clams, and squids, and the latter comprises the segmented worms, such as earthworms and leeches. These two groups have long been considered close relatives because of the common presence of trochophore larvae, but the annelids were considered closer to the arthropods because they are both segmented. Now, this is generally considered convergent evolution, owing to many morphological and genetic differences between the two phyla.

Among lesser phyla of invertebrates are the Hemichordata, or acorn worms, and the Chaetognatha, or arrow worms. Other phyla include Acoelomorpha, Brachiopoda, Bryozoa, Entoprocta, Phoronida, and Xenoturbellida.

Classification of Invertebrates

Invertebrates can be classified into several main categories, some of which are taxonomically obsolescent or debatable, but still used as terms of convenience.

- *Protozoa* (like the worms, an arbitrary grouping of convenience)

- Sponges (*Porifera*)

- Stinging jellyfish and corals (*Cnidaria*)

- Comb jellies (*Ctenophora*)

- Flatworms (*Platyhelminthes*)

- Round- or threadworms (*Nematoda*)

- segmented worms (*Annelida*)

- Insects, spiders, crabs and their kin (*Arthropoda*)

- Cuttlefish, snails, mussels and their kin (*Mollusca*)

- Starfish, sea-cucumbers and their kin (*Echinodermata*)

History

The earliest animal fossils appear to be those of invertebrates. 665-million-year-old fossils in the Trezona Formation at Trezona Bore, West Central Flinders, South Australia have been interpreted as being early sponges. Some paleontologists suggest that animals appeared much earlier, possibly as early as 1 billion years ago. Trace fossils such as tracks and burrows found in the Tonian era indicate the presence of triploblastic worms, like metazoans, roughly as large (about 5 mm wide) and complex as earthworms.

Around 453 MYA, animals began diversifying, and many of the important groups of invertebrates diverged from one another. Fossils of invertebrates are found in various types of sediment from the Phanerozoic. Fossils of invertebrates are commonly used in stratigraphy.

Classification

Carl Linnaeus divided these animals into only two groups, the Insecta and the now-obsolete Vermes (worms). Jean-Baptiste Lamarck, who was appointed to the position of "Curator of Insecta and Vermes" at the Muséum National d'Histoire Naturelle in 1793, both coined the term "invertebrate" to describe such animals, and divided the original two groups into ten, by splitting Arachnida and Crustacea from the Linnean Insecta, and Mollusca, Annelida, Cirripedia, Radiata, Coelenterata and Infusoria from the Linnean Vermes. They are now classified into over 30 phyla, from simple organisms such as sea sponges and flatworms to complex animals such as arthropods and molluscs.

Significance of the Group

Invertebrates are animals *without* a vertebral column. This has led to the conclusion that *inverte-brates* are a group that deviates from the normal, vertebrates. This has been said to be because researchers in the past, such as Lamarck, viewed vertebrates as a "standard": in Lamarck's theory of evolution, he believed that characteristics acquired through the evolutionary process involved not only survival, but also progression toward a "higher form", to which humans and vertebrates were closer than invertebrates were. Although goal-directed evolution has been abandoned, the distinction of invertebrates and vertebrates persists to this day, even though the grouping has been noted to be "hardly natural or even very sharp." Another reason cited for this continued distinction is that Lamarck created a precedent through his classifications which is now difficult to escape from. It is also possible that some humans believe that, they themselves being vertebrates, the group deserves more attention than invertebrates. In any event, in the 1968 edition of *Invertebrate Zool-ogy*, it is noted that "division of the Animal Kingdom into vertebrates and invertebrates is artificial and reflects human bias in favor of man's own relatives." The book also points out that the group lumps a vast number of species together, so that no one characteristic describes all invertebrates. In addition, some species included are only remotely related to one another, with some more related to vertebrates than other invertebrates.

In Research

For many centuries, invertebrates have been neglected by biologists, in favor of big vertebrates and "useful" or charismatic species. Invertebrate biology was not a major field of study until the work of Linnaeus and Lamarck in the 18th century. During the 20th century, invertebrate zoology became one of the major fields of natural sciences, with prominent discoveries in the fields of medicine, genetics, palaeontology, and ecology. The study of invertebrates has also benefited law enforcement, as arthropods, and especially insects, were discovered to be a source of information for forensic investigators.

Two of the most commonly studied model organisms nowadays are invertebrates: the fruit fly *Drosophila melanogaster* and the nematode *Caenorhabditis elegans*. They have long been the most intensively studied model organisms, and were among the first life-forms to be genetically sequenced. This was facilitated by the severely reduced state of their genomes, but many genes, introns, and linkages have been lost. Analysis of the starlet sea anemone genome has emphasised the importance of sponges, placozoans, and choanoflagellates, also being sequenced, in explaining the arrival of 1500 ancestral genes unique to animals. Invertebrates are also used by scientists in the field of aquatic biomonitoring to evaluate the effects of water pollution and climate change.

Insect

Insects are a class (Insecta) of hexapod invertebrates within the arthropod phylum that have a chitinous exoskeleton, a three-part body (head, thorax and abdomen), three pairs of jointed legs, compound eyes and one pair of antennae. They are the most diverse group of animals on the planet, including more than a million described species and representing more than half of all known living organisms. The number of extant species is estimated at between six and ten million, and potentially represent over 90% of the differing animal life forms on Earth. Insects may be found in nearly all environments, although only a small number of species reside in the oceans, a habitat dominated by another arthropod group, crustaceans.

The life cycles of insects vary but most hatch from eggs. Insect growth is constrained by the inelastic exoskeleton and development involves a series of molts. The immature stages can differ from the adults in structure, habit and habitat, and can include a passive pupal stage in those groups that undergo 4-stage metamorphosis. Insects that undergo 3-stage metamorphosis lack a pupal stage and adults develop through a series of nymphal stages. The higher level relationship of the Hexapoda is unclear. Fossilized insects of enormous size have been found from the Paleozoic Era, including giant dragonflies with wingspans of 55 to 70 cm (22–28 in). The most diverse insect groups appear to have coevolved with flowering plants.

Adult insects typically move about by walking, flying or sometimes swimming. As it allows for rapid yet stable movement, many insects adopt a tripedal gait in which they walk with their legs touching the ground in alternating triangles. Insects are the only invertebrates to have evolved flight. Many insects spend at least part of their lives under water, with larval adaptations that include gills, and some adult insects are aquatic and have adaptations for swimming. Some species, such as water striders, are capable of walking on the surface of water. Insects are mostly solitary, but some, such as certain bees, ants and termites, are social and live in large, well-organized colonies. Some insects, such as earwigs, show maternal care, guarding their eggs and young. Insects can communicate with each other in a variety of ways. Male moths can sense the pheromones of female moths over great distances. Other species communicate with sounds: crickets stridulate, or rub their wings together, to attract a mate and repel other males. Lampyridae in the beetle order communicate with light.

Humans regard certain insects as pests, and attempt to control them using insecticides and a host of other techniques. Some insects damage crops by feeding on sap, leaves or fruits. A few parasitic species are pathogenic. Some insects perform complex ecological roles; blow-flies, for example, help consume carrion but also spread diseases. Insect pollinators are essential to the life cycle of many flowering plant species on which most organisms, including humans, are at least partly dependent; without them, the terrestrial portion of the biosphere (including humans) would be devastated. Many other insects are considered ecologically beneficial as predators and a few provide direct economic benefit. Silkworms and bees have been used extensively by humans for the production of silk and honey, respectively. In some cultures, people eat the larvae or adults of certain insects.

Etymology

The word "insect" comes from the Latin word *insectum*, meaning "with a notched or divided body", or literally "cut into", from the neuter singular perfect passive participle of *insectare*, "to cut into, to cut up", from *in-* "into" and *secare* "to cut"; because insects appear "cut into" three sections. Pliny the Elder introduced the Latin designation as a loan-translation of the Greek word (*éntomos*) or "insect" (as in entomology), which was Aristotle's term for this class of life, also in reference to their "notched" bodies. "Insect" first appears documented in English in 1601 in Holland's translation of Pliny. Translations of Aristotle's term also form the usual word for "insect" in Welsh (trychfil, from *trychu* "to cut" and *mil*, "animal"), Serbo-Croatian (*zareznik*, from *rezati*, "to cut"), Russian (*nasekomoje*, from *seč/-sekat'*, "to cut"), etc.

Phylogeny and Evolution

The evolutionary relationship of insects to other animal groups remains unclear.

Although traditionally grouped with millipedes and centipedes—possibly on the basis of convergent adaptations to terrestrialisation—evidence has emerged favoring closer evolutionary ties with crustaceans. In the Pancrustacea theory, insects, together with Entognatha, Remipedia, and Cephalocarida, make up a natural clade labeled Miracrustacea.

A report in November 2014 unambiguously places the insects in one clade, with the crustaceans and myriapods, as the nearest sister clades. This study resolved insect phylogeny of all extant insect orders, and provides "a robust phylogenetic backbone tree and reliable time estimates of insect evolution."

Other terrestrial arthropods, such as centipedes, millipedes, scorpions, and spiders, are sometimes confused with insects since their body plans can appear similar, sharing (as do all arthropods) a jointed exoskeleton. However, upon closer examination, their features differ significantly; most noticeably, they do not have the six-legged characteristic of adult insects.

Evolution has produced enormous variety in insects.
Pictured are some of the possible shapes of antennae.

The higher-level phylogeny of the arthropods continues to be a matter of debate and research. In 2008, researchers at Tufts University uncovered what they believe is the world's oldest known full-body impression of a primitive flying insect, a 300 million-year-old specimen from the Carboniferous period. The oldest definitive insect fossil is the Devonian *Rhyniognatha hirsti*, from the 396-million-year-old Rhynie chert. It may have superficially resembled a modern-day silverfish insect. This species already possessed dicondylic mandibles (two articulations in the mandible), a feature associated with winged insects, suggesting that wings may already have evolved at this time. Thus, the first insects probably appeared earlier, in the Silurian period.

Four super radiations of insects have occurred: beetles (evolved about 300 million years ago), flies (evolved about 250 million years ago), and moths and wasps (evolved about 150 million years ago). These four groups account for the majority of described species. The flies and moths along with the fleas evolved from the Mecoptera.

The origins of insect flight remain obscure, since the earliest winged insects currently known appear to have been capable fliers. Some extinct insects had an additional pair of winglets attaching to the first segment of the thorax, for a total of three pairs. As of 2009, no evidence suggests the insects were a particularly successful group of animals before they evolved to have wings.

Late Carboniferous and Early Permian insect orders include both extant groups, their stem groups, and a number of Paleozoic groups, now extinct. During this era, some giant dragonfly-like forms reached wingspans of 55 to 70 cm (22 to 28 in), making them far larger than any living insect. This gigantism may have been due to higher atmospheric oxygen levels that allowed increased respiratory efficiency relative to today. The lack of flying vertebrates could have been another factor. Most extinct orders of insects developed during the Permian period that began around 270 million years ago. Many of the early groups became extinct during the Permian-Triassic extinction event, the largest mass extinction in the history of the Earth, around 252 million years ago.

The remarkably successful Hymenoptera appeared as long as 146 million years ago in the Cretaceous period, but achieved their wide diversity more recently in the Cenozoic era, which began 66 million years ago. A number of highly successful insect groups evolved in conjunction with flowering plants, a powerful illustration of coevolution.

Many modern insect genera developed during the Cenozoic. Insects from this period on are often found preserved in amber, often in perfect condition. The body plan, or morphology, of such specimens is thus easily compared with modern species. The study of fossilized insects is called paleoentomology.

Evolutionary Relationships

Insects are prey for a variety of organisms, including terrestrial vertebrates. The earliest vertebrates on land existed 400 million years ago and were large amphibious piscivores. Through gradual evolutionary change, insectivory was the next diet type to evolve.

Insects were among the earliest terrestrial herbivores and acted as major selection agents on plants. Plants evolved chemical defenses against this herbivory and the insects, in turn, evolved mechanisms to deal with plant toxins. Many insects make use of these toxins to protect themselves from their predators. Such insects often advertise their toxicity using warning colors. This successful evolutionary pattern has also been used by mimics. Over time, this has led to complex groups of coevolved species. Conversely, some interactions between plants and insects, like pollination, are beneficial to both organisms. Coevolution has led to the development of very specific mutualisms in such systems.

Taxonomy

Cladogram of living insect groups, with numbers of species in each group. The Apterygota, Palaeoptera, and Exopterygota are possibly paraphyletic groups.

Traditional morphology-based or appearance-based systematics have usually given the Hexapoda the rank of superclass, and identified four groups within it: insects (Ectognatha), springtails (Collembola), Protura, and Diplura, the latter three being grouped together as the Entognatha on the basis of internalized mouth parts. Supraordinal relationships have undergone numerous changes with the advent of methods based on evolutionary history and genetic data. A recent theory is that the Hexapoda are polyphyletic (where the last common ancestor was not a member of the group), with the entognath classes having separate evolutionary histories from the Insecta. Many of the traditional appearance-based taxa have been shown to be paraphyletic, so rather than using ranks like subclass, superorder, and infraorder, it has proved better to use monophyletic groupings (in which the last common ancestor is a member of the group). The following represents the best-supported monophyletic groupings for the Insecta.

Insects can be divided into two groups historically treated as subclasses: wingless insects, known as Apterygota, and winged insects, known as Pterygota. The Apterygota consist of the primitively wingless order of the silverfish (Zygentoma). Archaeognatha make up the Mono-condylia based on the shape of their mandibles, while Zygentoma and Pterygota are grouped together as Dicondylia. The Zygentoma themselves possibly are not monophyletic, with the family Lepidotrichidae being a sister group to the Dicondylia (Pterygota and the remaining Zygentoma).

Paleoptera and Neoptera are the winged orders of insects differentiated by the presence of hardened body parts called sclerites, and in the Neoptera, muscles that allow their wings to fold flatly over the abdomen. Neoptera can further be divided into incomplete metamorphosis-based (Polyneoptera and Paraneoptera) and complete metamorphosis-based groups. It has proved difficult to clarify the relationships between the orders in Polyneoptera because of constant new findings calling for revision of the taxa. For example, the Paraneoptera have turned out to be more closely related to the Endopterygota than to the rest of the Exopterygota. The recent molecular finding that the traditional louse orders Mallophaga and Anoplura are derived from within Psocoptera has led to the new taxon Psocodea. Phasmatodea and Embiidina have been suggested to form the Eukinolabia. Mantodea, Blattodea, and Isoptera are thought to form a monophyletic group termed Dictyoptera.

The Exopterygota likely are paraphyletic in regard to the Endopterygota. Matters that have incurred controversy include Strepsiptera and Diptera grouped together as Halteria based on a reduction of one of the wing pairs – a position not well-supported in the entomological community. The Neuropterida are often lumped or split on the whims of the taxonomist. Fleas are now thought to be closely related to boreid mecopterans. Many questions remain in the basal relationships amongst endopterygote orders, particularly the Hymenoptera.

The study of the classification or taxonomy of any insect is called systematic entomology. If one works with a more specific order or even a family, the term may also be made specific to that order or family, for example systematic dipterology.

Diversity

Though the true dimensions of species diversity remain uncertain, estimates range from 2.6–7.8 million species with a mean of 5.5 million. This probably represents less than 20% of all species on Earth, and with only about 20,000 new species of all organisms being described each year, most species likely will remain undescribed for many years unless species descriptions increase in rate. About 850,000–1,000,000 of all described species are insects. Of the 24 orders of insects, four dominate in terms of numbers of described species, with at least 670,000 species included in Coleoptera, Diptera, Hymenoptera and Lepidoptera.

Comparison of the estimated number of species in the four most speciose insect orders			
	Described species	**Average description rate (species per year)**	**Publication effort**
Coleoptera	300,000–400,000	2308	0.01
Lepidoptera	180,000	642	0.03
Diptera	90,000–150,000	1048	0.04
Hymenoptera	100,000–150,000	1196	0.02

A 2015 study estimated the number of beetles at 0.9–2.1 million with a mean of 1.5 million.

Morphology and Physiology

External

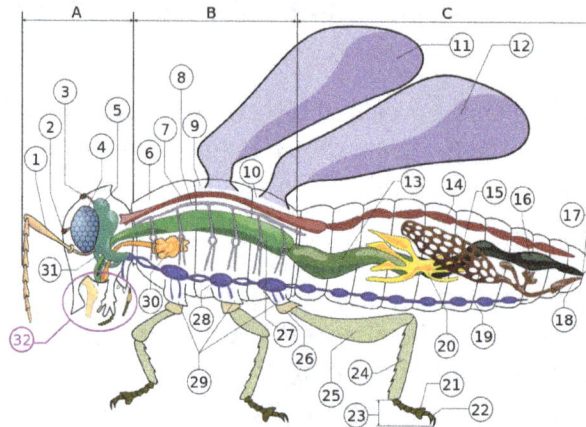

Insect morphology
A- Head B- Thorax C- Abdomen

1. antenna	9. mesothorax	17. anus	25. femur
2. ocelli (lower)	10. metathorax	18. oviduct	26. trochanter
3. ocelli (upper)	11. forewing	19. nerve chord (abdominal ganglia)	27. fore-gut (crop, gizzard)
4. compound eye	12. hindwing		28. thoracic ganglion
5. brain (cerebral ganglia)	13. mid-gut (stomach)	20. Malpighian tubes	29. coxa
6. prothorax	14. dorsal tube (Heart)	21. tarsal pads	30. salivary gland
7. dorsal blood vessel	15. ovary	22. claws	31. subesophageal ganglion
8. tracheal tubes (trunk with spiracle)	16. hind-gut (intestine, rectum & anus)	23. tarsus	32. mouthparts
		24. tibia	

Insects have segmented bodies supported by exoskeletons, the hard outer covering made mostly of chitin. The segments of the body are organized into three distinctive but interconnected units, or tagmata: a head, a thorax and an abdomen. The head supports a pair of sensory antennae, a pair of compound eyes, and, if present, one to three simple eyes (or ocelli) and three sets of variously modified appendages that form the mouthparts. The thorax has six segmented legs—one pair each for the prothorax, mesothorax and the metathorax segments making up the thorax—and, none, two or four wings. The abdomen consists of eleven segments, though in a few species of insects, these segments may be fused together or reduced in size. The abdomen also contains most of the digestive, respiratory, excretory and reproductive internal structures. Considerable variation and many adaptations in the body parts of insects occur, especially wings, legs, antenna and mouthparts.

Segmentation

The head is enclosed in a hard, heavily sclerotized, unsegmented, exoskeletal head capsule, or epicranium, which contains most of the sensing organs, including the antennae, ocellus or eyes, and the mouthparts. Of all the insect orders, Orthoptera displays the most features found in other insects, including the sutures and sclerites. Here, the vertex, or the apex (dorsal region), is situated between the compound eyes for insects with a hypognathous and opisthognathous head. In prognathous insects, the vertex is not found between the compound eyes, but rather, where the ocelli are normally. This is because the primary axis of the head is rotated 90° to become parallel to the primary axis of the body. In some species, this region is modified and assumes a different name.

The thorax is a tagma composed of three sections, the prothorax, mesothorax and the metathorax. The anterior segment, closest to the head, is the prothorax, with the major features being the first pair of legs and the pronotum. The middle segment is the mesothorax, with the major features being the second pair of legs and the anterior wings. The third and most posterior segment, abutting the abdomen, is the metathorax, which features the third pair of legs and the posterior wings. Each segment is dilineated by an intersegmental suture. Each segment has four basic regions. The dorsal surface is called the tergum (or *notum*) to distinguish it from the abdominal terga. The two lateral regions are called the pleura (singular: pleuron) and the ventral aspect is called the sternum. In turn, the notum of the prothorax is called the pronotum, the notum for the mesothorax is called the mesonotum and the notum for the metathorax is called the metanotum. Continuing with this logic, the mesopleura and metapleura, as well as the mesosternum and metasternum, are used.

The abdomen is the largest tagma of the insect, which typically consists of 11–12 segments and is less strongly sclerotized than the head or thorax. Each segment of the abdomen is represented by a sclerotized tergum and sternum. Terga are separated from each other and from the adjacent sterna or pleura by membranes. Spiracles are located in the pleural area. Variation of this ground plan includes the fusion of terga or terga and sterna to form continuous dorsal or ventral shields or a conical tube. Some insects bear a sclerite in the pleural area called a laterotergite. Ventral sclerites are sometimes called laterosternites. During the embryonic stage of many insects and the postembryonic stage of primitive insects, 11 abdominal segments are present. In modern insects there is a tendency toward reduction in the number of the abdominal segments, but the primitive number of 11 is maintained during embryogenesis. Variation in abdominal segment number is considerable. If the Apterygota are considered to be indicative of the ground plan for pterygotes, confusion reigns: adult Protura have 12 segments, Collembola have 6. The orthopteran family Acrididae has 11 segments, and a fossil specimen of Zoraptera has a 10-segmented abdomen.

Exoskeleton

The insect outer skeleton, the cuticle, is made up of two layers: the epicuticle, which is a thin and waxy water resistant outer layer and contains no chitin, and a lower layer called the procuticle. The procuticle is chitinous and much thicker than the epicuticle and has two layers: an outer layer known as the exocuticle and an inner layer known as the endocuticle. The tough and flexible endocuticle is built from numerous layers of fibrous chitin and proteins, criss-crossing each other in a sandwich pattern, while the exocuticle is rigid and hardened. The exocuticle is greatly reduced in many soft-bodied insects (e.g., caterpillars), especially during their larval stages.

Insects are the only invertebrates to have developed active flight capability, and this has played an important role in their success. Their muscles are able to contract multiple times for each single nerve impulse, allowing the wings to beat faster than would ordinarily be possible. Having their muscles attached to their exoskeletons is more efficient and allows more muscle connections; crustaceans also use the same method, though all spiders use hydraulic pressure to extend their legs, a system inherited from their pre-arthropod ancestors. Unlike insects, though, most aquatic crustaceans are biomineralized with calcium carbonate extracted from the water.

Internal

Nervous System

The nervous system of an insect can be divided into a brain and a ventral nerve cord. The head capsule is made up of six fused segments, each with either a pair of ganglia, or a cluster of nerve cells outside of the brain. The first three pairs of ganglia are fused into the brain, while the three following pairs are fused into a structure of three pairs of ganglia under the insect's esophagus, called the subesophageal ganglion.

The thoracic segments have one ganglion on each side, which are connected into a pair, one pair per segment. This arrangement is also seen in the abdomen but only in the first eight segments. Many species of insects have reduced numbers of ganglia due to fusion or reduction. Some cockroaches have just six ganglia in the abdomen, whereas the wasp *Vespa crabro* has only two in the thorax and three in the abdomen. Some insects, like the house fly *Musca domestica*, have all the body ganglia fused into a single large thoracic ganglion.

At least a few insects have nociceptors, cells that detect and transmit signals responsible for the sensation of pain. This was discovered in 2003 by studying the variation in reactions of larvae of the common fruitfly Drosophila to the touch of a heated probe and an unheated one. The larvae reacted to the touch of the heated probe with a stereotypical rolling behavior that was not exhibited when the larvae were touched by the unheated probe. Although nociception has been demonstrated in insects, there is no consensus that insects feel pain consciously insects are capable of learning.

Digestive System

An insect uses its digestive system to extract nutrients and other substances from the food it consumes. Most of this food is ingested in the form of macromolecules and other complex substances like proteins, polysaccharides, fats and nucleic acids. These macromolecules must be broken down by catabolic reactions into smaller molecules like amino acids and simple sugars before being used by cells of the body for energy, growth, or reproduction. This break-down process is known as digestion.

The main structure of an insect's digestive system is a long enclosed tube called the alimentary canal, which runs lengthwise through the body. The alimentary canal directs food unidirectionally from the mouth to the anus. It has three sections, each of which performs a different process of digestion. In addition to the alimentary canal, insects also have paired salivary glands and salivary reservoirs. These structures usually reside in the thorax, adjacent to the foregut.

The salivary glands in an insect's mouth produce saliva. The salivary ducts lead from the glands to the reservoirs and then forward through the head to an opening called the salivarium, located behind the hypopharynx. By moving its mouthparts the insect can mix its food with saliva. The mixture of saliva and food then travels through the salivary tubes into the mouth, where it begins to break down. Some insects, like flies, have extra-oral digestion. Insects using extra-oral digestion expel digestive enzymes onto their food to break it down. This strategy allows insects to extract a significant proportion of the available nutrients from the food source. The gut is where almost all of insects' digestion takes place. It can be divided into the foregut, midgut and hindgut.

Foregut

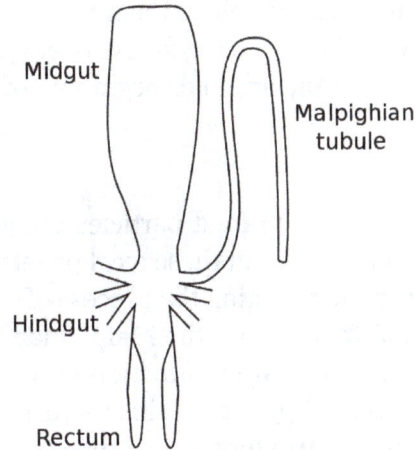

Stylized diagram of insect digestive tract showing malpighian tubule, from an insect of the order Orthoptera.

The first section of the alimentary canal is the foregut, or stomodaeum. The foregut is lined with a cuticular lining made of chitin and proteins as protection from tough food. The foregut includes the buccal cavity (mouth), pharynx, esophagus and crop and proventriculus (any part may be highly modified) which both store food and signify when to continue passing onward to the midgut.

Digestion starts in buccal cavity (mouth) as partially chewed food is broken down by saliva from the salivary glands. As the salivary glands produce fluid and carbohydrate-digesting enzymes (mostly amylases), strong muscles in the pharynx pump fluid into the buccal cavity, lubricating the food like the salivarium does, and helping blood feeders, and xylem and phloem feeders.

From there, the pharynx passes food to the esophagus, which could be just a simple tube passing it on to the crop and proventriculus, and then onward to the midgut, as in most insects. Alternately, the foregut may expand into a very enlarged crop and proventriculus, or the crop could just be a diverticulum, or fluid-filled structure, as in some Diptera species.

Bumblebee defecating

Midgut

Once food leaves the crop, it passes to the midgut, also known as the mesenteron, where the majority of digestion takes place. Microscopic projections from the midgut wall, called microvilli, increase the surface area of the wall and allow more nutrients to be absorbed; they tend to be close to the origin of the midgut. In some insects, the role of the microvilli and where they are located may vary. For example, specialized microvilli producing digestive enzymes may more likely be near the end of the midgut, and absorption near the origin or beginning of the midgut.

Hindgut

In the hindgut, or proctodaeum, undigested food particles are joined by uric acid to form fecal pellets. The rectum absorbs 90% of the water in these fecal pellets, and the dry pellet is then eliminated through the anus (element 17), completing the process of digestion. The uric acid is formed using hemolymph waste products diffused from the Malpighian tubules (element 20). It is then emptied directly into the alimentary canal, at the junction between the midgut and hindgut. The number of Malpighian tubules possessed by a given insect varies between species, ranging from only two tubules in some insects to over 100 tubules in others.

Reproductive System

The reproductive system of female insects consist of a pair of ovaries, accessory glands, one or more spermathecae, and ducts connecting these parts. The ovaries are made up of a number of egg tubes, called ovarioles, which vary in size and number by species. The number of eggs that the insect is able to make vary by the number of ovarioles with the rate that eggs can develop being also influenced by ovariole design. Female insects are able make eggs, receive and store sperm, manipulate sperm from different males, and lay eggs. Accessory glands or glandular parts of the oviducts produce a variety of substances for sperm maintenance, transport and fertilization, as well as for protection of eggs. They can produce glue and protective substances for coating eggs or tough coverings for a batch of eggs called oothecae. Spermathecae are tubes or sacs in which sperm can be stored between the time of mating and the time an egg is fertilized.

For males, the reproductive system is the testis, suspended in the body cavity by tracheae and the fat body. Most male insects have a pair of testes, inside of which are sperm tubes or follicles that are enclosed within a membranous sac. The follicles connect to the vas deferens by the vas efferens, and the two tubular vasa deferentia connect to a median ejaculatory duct that leads to the outside. A portion of the vas deferens is often enlarged to form the seminal vesicle, which stores the sperm before they are discharged into the female. The seminal vesicles have glandular linings that secrete nutrients for nourishment and maintenance of the sperm. The ejaculatory duct is derived from an invagination of the epidermal cells during development and, as a result, has a cuticular lining. The terminal portion of the ejaculatory duct may be sclerotized to form the intromittent organ, the aedeagus. The remainder of the male reproductive system is derived from embryonic mesoderm, except for the germ cells, or spermatogonia, which descend from the primordial pole cells very early during embryogenesis.

Respiratory System

The tube-like heart (green) of the mosquito *Anopheles gambiae* extends horizontally across the body, interlinked with the diamond-shaped wing muscles (also green) and surrounded by pericardial cells (red). Blue depicts cell nuclei.

Insect respiration is accomplished without lungs. Instead, the insect respiratory system uses a system of internal tubes and sacs through which gases either diffuse or are actively pumped, delivering oxygen directly to tissues that need it via their trachea. Since oxygen is delivered directly, the circulatory system is not used to carry oxygen, and is therefore greatly reduced. The insect circulatory system has no veins or arteries, and instead consists of little more than a single, perforated dorsal tube which pulses peristaltically. Toward the thorax, the dorsal tube divides into chambers and acts like the insect's heart. The opposite end of the dorsal tube is like the aorta of the insect circulating the hemolymph, arthropods' fluid analog of blood, inside the body cavity. Air is taken in through openings on the sides of the abdomen called spiracles.

The respiratory system is an important factor that limits the size of insects. As insects get bigger, this type of oxygen transport gets less efficient and thus the heaviest insect currently weighs less than 100 g. However, with increased atmospheric oxygen levels, as happened in the late Paleozoic, larger insects were possible, such as dragonflies with wingspans of more than two feet.

There are many different patterns of gas exchange demonstrated by different groups of insects. Gas exchange patterns in insects can range from continuous and diffusive ventilation, to discontinuous gas exchange. During continuous gas exchange, oxygen is taken in and carbon dioxide is released in a continuous cycle. In discontinuous gas exchange, however, the insect takes in oxygen while it is active and small amounts of carbon dioxide are released when the insect is at rest. Diffusive ventilation is simply a form of continuous gas exchange that occurs by diffusion rather than physically taking in the oxygen. Some species of insect that are submerged also have adaptations to aid in respiration. As larvae, many insects have gills that can extract oxygen dissolved in water, while others need to rise to the water surface to replenish air supplies which may be held or trapped in special structures.

Circulatory System

The insect circulatory system utilizes hemolymph, a tissue analogous to blood that circulates in the interior of the insect body, while remaining in direct contact with the animal's tissues. It is composed of plasma in which hemocytes are suspended. In addition to hemocytes, the plasma also contains many chemicals. It is also the major tissue type of the open circulatory system of arthropods, characteristic of spiders, crustaceans and insects.

Reproduction and Development

The majority of insects hatch from eggs. The fertilization and development takes place inside the egg, enclosed by a shell (chorion) that consists of maternal tissue. In contrast to eggs of other arthropods, most insect eggs are drought resistant. This is because inside the chorion two additional membranes develop from embryonic tissue, the amnion and the serosa. This serosa secretes a cuticle rich in chitin that protects the embryo against desiccation. In Schizophora however the serosa does not develop, but these flies lay their eggs in damp places, such as rotting matter. Some species of insects, like the cockroach *Blaptica dubia*, as well as juvenile aphids and tsetse flies, are ovoviviparous. The eggs of ovoviviparous animals develop entirely inside the female, and then hatch immediately upon being laid. Some other species, such as those in the genus of cockroaches known as *Diploptera*, are viviparous, and thus gestate inside the mother and are born alive.Some insects, like parasitic wasps, show polyembryony, where a single fertilized egg divides into many and in some cases thousands of separate embryos. Insects may be *univoltine, bivoltine* or *multivoltine,* i.e. they may have one, two or many broods (generations) in a year.

A pair of *Simosyrphus grandicornis* hoverflies mating in flight.

A pair of *grasshoppers* mating.

The different forms of the male (top) and female (bottom)
tussock moth *Orgyia recens* is an example of sexual dimorphism in insects.

Other developmental and reproductive variations include haplodiploidy, polymorphism, paedomorphosis or peramorphosis, sexual dimorphism, parthenogenesis and more rarely hermaphroditism. In haplodiploidy, which is a type of sex-determination system, the offspring's sex is determined by the number of sets of chromosomes an individual receives. This system is typical in bees and wasps. Polymorphism is where a species may have different *morphs* or *forms*, as in the oblong winged katydid, which has four different varieties: green, pink and yellow or tan. Some insects may retain phenotypes that are normally only seen in juveniles; this is called paedomorphosis. In peramorphosis, an opposite sort of phenomenon, insects take on previously unseen traits after they have matured into adults. Many insects display sexual dimorphism, in which males and females have notably different appearances, such as the moth *Orgyia recens* as an exemplar of sexual dimorphism in insects.

Some insects use parthenogenesis, a process in which the female can reproduce and give birth without having the eggs fertilized by a male. Many aphids undergo a form of parthenogenesis, called cyclical parthenogenesis, in which they alternate between one or many generations of asexual and sexual reproduction. In summer, aphids are generally female and parthenogenetic; in the autumn, males may be produced for sexual reproduction. Other insects produced by parthenogenesis are bees, wasps and ants, in which they spawn males. However, overall, most individuals are female, which are produced by fertilization. The males are haploid and the females are diploid. More rarely, some insects display hermaphroditism, in which a given individual has both male and female reproductive organs.

Insect life-histories show adaptations to withstand cold and dry conditions. Some temperate region insects are capable of activity during winter, while some others migrate to a warmer climate or go into a state of torpor. Still other insects have evolved mechanisms of diapause that allow eggs or pupae to survive these conditions.

Metamorphosis

Metamorphosis in insects is the biological process of development all insects must undergo. There are two forms of metamorphosis: incomplete metamorphosis and complete metamorphosis.

Incomplete Metamorphosis

Hemimetabolous insects, those with incomplete metamorphosis, change gradually by undergoing a series of molts. An insect molts when it outgrows its exoskeleton, which does not stretch and would otherwise restrict the insect's growth. The molting process begins as the insect's epidermis secretes a new epicuticle inside the old one. After this new epicuticle is secreted, the epidermis releases a mixture of enzymes that digests the endocuticle and thus detaches the old cuticle. When this stage is complete, the insect makes its body swell by taking in a large quantity of water or air, which makes the old cuticle split along predefined weaknesses where the old exocuticle was thinnest.

This southern hawker dragonfly molts its exoskeleton several times
during its life as a nymph; shown is the final molt to become a winged adult (eclosion).

Immature insects that go through incomplete metamorphosis are called nymphs or in the case of dragonflies and damselflies, also naiads. Nymphs are similar in form to the adult except for the presence of wings, which are not developed until adulthood. With each molt, nymphs grow larger and become more similar in appearance to adult insects.

Complete Metamorphosis

Gulf Fritillary Life Cycle

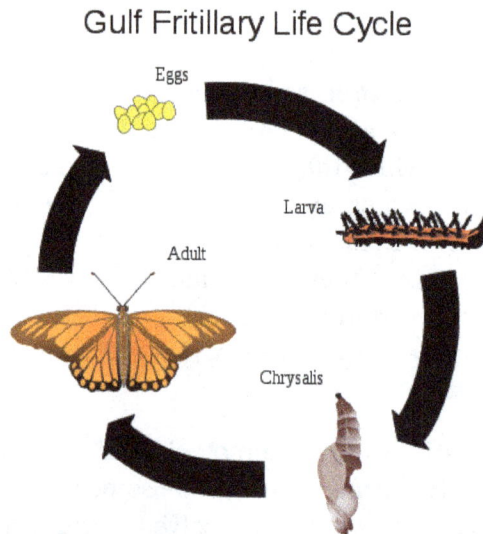

Gulf fritillary life cycle, an example of holometabolism.

Holometabolism, or complete metamorphosis, is where the insect changes in four stages, an egg or embryo, a larva, a pupa and the adult or imago. In these species, an egg hatches to produce a larva, which is generally worm-like in form. This worm-like form can be one of several varieties: eruciform (caterpillar-like), scarabaeiform (grub-like), campodeiform (elongated, flattened and active), elateriform (wireworm-like) or vermiform (maggot-like). The larva grows and eventually becomes a pupa, a stage marked by reduced movement and often sealed within a cocoon. There are three types of pupae: obtect, exarate or coarctate. Obtect pupae are compact, with the legs and other appendages enclosed. Exarate pupae have their legs and other appendages free and extended. Coarctate pupae develop inside the larval skin. Insects undergo considerable change in form during the pupal stage, and emerge as adults. Butterflies are a well-known example of insects that undergo complete metamorphosis, although most insects use this life cycle. Some insects have evolved this system to hypermetamorphosis.

Some of the oldest and most successful insect groups, such Endopterygota, use a system of complete metamorphosis. Complete metamorphosis is unique to a group of certain insect orders including Diptera, Lepidoptera and Hymenoptera. This form of development is exclusive and not seen in any other arthropods.

Senses and Communication

Many insects possess very sensitive and, or specialized organs of perception. Some insects such as bees can perceive ultraviolet wavelengths, or detect polarized light, while the antennae of male moths can detect the pheromones of female moths over distances of many kilometers. The yellow paper wasp (*Polistes versicolor*) is known for its wagging movements as a form of communication within the colony; it can waggle with a frequency of 10.6±2.1 Hz (n=190). These wagging movements can signal the arrival of new material into the nest and aggression between workers can be used to stimulate others to increase foraging expeditions. There is a pronounced tendency for there to be a trade-off between visual acuity and chemical or tactile acuity, such that most insects with well-developed eyes have reduced or simple antennae, and vice versa. There are a variety of different mechanisms by which insects perceive sound, while the patterns are not universal, insects can generally hear sound if they can produce it. Different insect species can have varying hearing, though most insects can hear only a narrow range of frequencies related to the frequency of the sounds they can produce. Mosquitoes have been found to hear up to 2 kHz, and some grasshoppers can hear up to 50 kHz. Certain predatory and parasitic insects can detect the characteristic sounds made by their prey or hosts, respectively. For instance, some nocturnal moths can perceive the ultrasonic emissions of bats, which helps them avoid predation. Insects that feed on blood have special sensory structures that can detect infrared emissions, and use them to home in on their hosts.

Some insects display a rudimentary sense of numbers, such as the solitary wasps that prey upon a single species. The mother wasp lays her eggs in individual cells and provides each egg with a number of live caterpillars on which the young feed when hatched. Some species of wasp always provide five, others twelve, and others as high as twenty-four caterpillars per cell. The number of caterpillars is different among species, but always the same for each sex of larva. The male solitary wasp in the genus *Eumenes* is smaller than the female, so the mother of one species supplies him with only five caterpillars; the larger female receives ten caterpillars in her cell.

Light Production and Vision

Most insects have compound eyes and two antennae.

A few insects, such as members of the families Poduridae and Onychiuridae (Collembola), Mycetophilidae (Diptera) and the beetle families Lampyridae, Phengodidae, Elateridae and Staphylinidae are bioluminescent. The most familiar group are the fireflies, beetles of the family Lampyridae.

Some species are able to control this light generation to produce flashes. The function varies with some species using them to attract mates, while others use them to lure prey. Cave dwelling larvae of *Arachnocampa* (Mycetophilidae, fungus gnats) glow to lure small flying insects into sticky strands of silk. Some fireflies of the genus *Photuris* mimic the flashing of female *Photinus* species to attract males of that species, which are then captured and devoured. The colors of emitted light vary from dull blue (*Orfelia fultoni*, Mycetophilidae) to the familiar greens and the rare reds (*Phrixothrix tiemanni*, Phengodidae).

Most insects, except some species of cave crickets, are able to perceive light and dark. Many species have acute vision capable of detecting minute movements. The eyes may include simple eyes or ocelli as well as compound eyes of varying sizes. Many species are able to detect light in the infrared, ultraviolet and the visible light wavelengths. Color vision has been demonstrated in many species and phylogenetic analysis suggests that UV-green-blue trichromacy existed from at least the Devonian period between 416 and 359 million years ago.

Sound Production and Hearing

Insects were the earliest organisms to produce and sense sounds. Insects make sounds mostly by mechanical action of appendages. In grasshoppers and crickets, this is achieved by stridulation. Cicadas make the loudest sounds among the insects by producing and amplifying sounds with special modifications to their body and musculature. The African cicada *Brevisana brevis* has been measured at 106.7 decibels at a distance of 50 cm (20 in). Some insects, such as the *Helicoverpa zea* moths, hawk moths and Hedylid butterflies, can hear ultrasound and take evasive action when they sense that they have been detected by bats. Some moths produce ultrasonic clicks that were once thought to have a role in jamming bat echolocation. The ultrasonic clicks were subsequently found to be produced mostly by unpalatable moths to warn bats, just as warning colorations are used against predators that hunt by sight. Some otherwise palatable moths have evolved to mimic these calls. More recently, the claim that some moths can jam bat sonar has been revisited. Ultrasonic recording and high-speed infrared videography of bat-moth interactions suggest the palatable tiger moth really does defend against attacking big brown bats using ultrasonic clicks that jam bat sonar.

Very low sounds are also produced in various species of Coleoptera, Hymenoptera, Lepidoptera, Mantodea and Neuroptera. These low sounds are simply the sounds made by the insect's movement. Through microscopic stridulatory structures located on the insect's muscles and joints, the normal sounds of the insect moving are amplified and can be used to warn or communicate with other insects. Most sound-making insects also have tympanal organs that can perceive airborne sounds. Some species in Hemiptera, such as the corixids (water boatmen), are known to communicate via underwater sounds. Most insects are also able to sense vibrations transmitted through surfaces.

Communication using surface-borne vibrational signals is more widespread among insects because of size constraints in producing air-borne sounds. Insects cannot effectively produce low-frequency sounds, and high-frequency sounds tend to disperse more in a dense environment (such as foliage), so insects living in such environments communicate primarily using substrate-borne vibrations. The mechanisms of production of vibrational signals are just as diverse as those for producing sound in insects.

Some species use vibrations for communicating within members of the same species, such as to attract mates as in the songs of the shield bug *Nezara viridula*. Vibrations can also be used to communicate between entirely different species; lycaenid (gossamer-winged butterfly) caterpillars which are myrmecophilous (living in a mutualistic association with ants) communicate with ants in this way. The Madagascar hissing cockroach has the ability to press air through its spiracles to make a hissing noise as a sign of aggression; the death's-head hawkmoth makes a squeaking noise by forcing air out of their pharynx when agitated, which may also reduce aggressive worker honey bee behavior when the two are in close proximity.

Chemical Communication

Chemical communications in animals rely on a variety of aspects including taste and smell. Chemoreception is the physiological response of a sense organ (i.e. taste or smell) to a chemical stimulus where the chemicals act as signals to regulate the state or activity of a cell. A semiochemical is a message-carrying chemical that is meant to attract, repel, and convey information. Types of semiochemicals include pheromones and kairomones. One example is the butterfly *Phengaris arion* which uses chemical signals as a form of mimicry to aid in predation.

In addition to the use of sound for communication, a wide range of insects have evolved chemical means for communication. These chemicals, termed semiochemicals, are often derived from plant metabolites include those meant to attract, repel and provide other kinds of information. Pheromones, a type of semiochemical, are used for attracting mates of the opposite sex, for aggregating conspecific individuals of both sexes, for deterring other individuals from approaching, to mark a trail, and to trigger aggression in nearby individuals. Allomonea benefit their producer by the effect they have upon the receiver. Kairomones benefit their receiver instead of their producer. Synomones benefit the producer and the receiver. While some chemicals are targeted at individuals of the same species, others are used for communication across species. The use of scents is especially well known to have developed in social insects.

Social Behavior

A cathedral mound created by termites (Isoptera).

Social insects, such as termites, ants and many bees and wasps, are the most familiar species of eusocial animal. They live together in large well-organized colonies that may be so tightly integrated and genetically similar that the colonies of some species are sometimes considered superorganisms. It is sometimes argued that the various species of honey bee are the only invertebrates (and indeed one of the few non-human groups) to have evolved a system of abstract symbolic communication where a behavior is used to *represent* and convey specific information about something in the environment. In this communication system, called dance language, the angle at which a bee dances represents a direction relative to the sun, and the length of the dance represents the distance to be flown. Though perhaps not as advanced as honey bees, bumblebees also potentially have some social communication behaviors. *Bombus terrestris*, for example, exhibit a faster learning curve for visiting unfamiliar, yet rewarding flowers, when they can see a conspecific foraging on the same species.

Only insects which live in nests or colonies demonstrate any true capacity for fine-scale spatial orientation or homing. This can allow an insect to return unerringly to a single hole a few millimeters in diameter among thousands of apparently identical holes clustered together, after a trip of up to several kilometers' distance. In a phenomenon known as philopatry, insects that hibernate have shown the ability to recall a specific location up to a year after last viewing the area of interest. A few insects seasonally migrate large distances between different geographic regions (e.g., the overwintering areas of the monarch butterfly).

Care of Young

The eusocial insects build nest, guard eggs, and provide food for offspring full-time. Most insects, however, lead short lives as adults, and rarely interact with one another except to mate or compete for mates. A small number exhibit some form of parental care, where they will at least guard their eggs, and sometimes continue guarding their offspring until adulthood, and possibly even feeding them. Another simple form of parental care is to construct a nest (a burrow or an actual construction, either of which may be simple or complex), store provisions in it, and lay an egg upon those provisions. The adult does not contact the growing offspring, but it nonetheless does provide food. This sort of care is typical for most species of bees and various types of wasps.

Locomotion

Flight

Insects are the only group of invertebrates to have developed flight. The evolution of insect wings has been a subject of debate. Some entomologists suggest that the wings are from paranotal lobes, or extensions from the insect's exoskeleton called the nota, called the *paranotal theory*. Other theories are based on a pleural origin. These theories include suggestions that wings originated from modified gills, spiracular flaps or as from an appendage of the epicoxa. The *epicoxal theory* suggests the insect wings are modified epicoxal exites, a modified appendage at the base of the legs or coxa. In the Carboniferous age, some of the *Meganeura* dragonflies had as much as a 50 cm (20 in) wide wingspan. The appearance of gigantic insects has been found to be consistent with high atmospheric oxygen. The respiratory system of insects constrains their size, however the high oxygen in the atmosphere allowed larger sizes.

The largest flying insects today are much smaller and include several moth species such as the Atlas moth and the white witch (*Thysania agrippina*).

Unlike birds, many small insects are swept along by the prevailing winds although many of the larger insects are known to make migrations. Aphids are known to be transported long distances by low-level jet streams. As such, fine line patterns associated with converging winds within weather radar imagery, like the WSR-88D radar network, often represent large groups of insects.

White-lined sphinx moth feeding in flight

Insect flight has been a topic of great interest in aerodynamics due partly to the inability of steady-state theories to explain the lift generated by the tiny wings of insects. But insect wings are in motion, with flapping and vibrations, resulting in churning and eddies, and the misconception that physics says "bumblebees can't fly" persisted throughout most of the twentieth century.

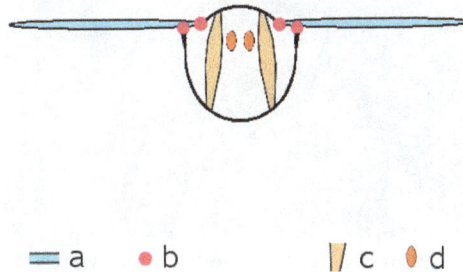

Basic motion of the insect wing in insect with an indirect flight mechanism scheme of dorsoventral cut through a thorax segment with a wings b joints c dorsoventral muscles d longitudinal muscles.

Walking

Many adult insects use six legs for walking and have adopted a tripedal gait. The tripedal gait allows for rapid walking while always having a stable stance and has been studied extensively in cockroaches and ants. The legs are used in alternate triangles touching the ground. For the first step, the middle right leg and the front and rear left legs are in contact with the ground and move the insect forward, while the front and rear right leg and the middle left leg are lifted and moved forward to a new position. When they touch the ground to form a new stable triangle the other legs can be lifted and brought forward in turn and so on. The purest form of the tripedal gait is seen in insects moving at high speeds. However, this type of locomotion is not rigid and insects can adapt a variety of gaits. For example, when moving slowly, turning, avoiding obstacles, climbing or slippery surfaces, four (tetrapod) or more feet (wave-gait) may be touching the ground. Insects can also adapt their gait to cope with the loss of one or more limbs.

Cockroaches are among the fastest insect runners and, at full speed, adopt a bipedal run to reach a high velocity in proportion to their body size. As cockroaches move very quickly, they need to be video recorded at several hundred frames per second to reveal their gait. More sedate locomotion is seen in the stick insects or walking sticks (Phasmatodea). A few insects have evolved to walk on the surface of the water, especially members of the Gerridae family, commonly known as water striders. A few species of ocean-skaters in the genus *Halobates* even live on the surface of open oceans, a habitat that has few insect species.

Use in Robotics

Insect walking is of particular interest as an alternative form of locomotion in robots. The study of insects and bipeds has a significant impact on possible robotic methods of transport. This may allow new robots to be designed that can traverse terrain that robots with wheels may be unable to handle.

Swimming

The backswimmer *Notonecta glauca* underwater,
showing its paddle-like hindleg adaptation.

A large number of insects live either part or the whole of their lives underwater. In many of the more primitive orders of insect, the immature stages are spent in an aquatic environment. Some groups of insects, like certain water beetles, have aquatic adults as well.

Many of these species have adaptations to help in under-water locomotion. Water beetles and water bugs have legs adapted into paddle-like structures. Dragonfly naiads use jet propulsion, forcibly expelling water out of their rectal chamber. Some species like the water striders are capable of walking on the surface of water. They can do this because their claws are not at the tips of the legs as in most insects, but recessed in a special groove further up the leg; this prevents the claws from piercing the water's surface film. Other insects such as the Rove beetle *Stenus* are known to emit pygidial gland secretions that reduce surface tension making it possible for them to move on the surface of water by Marangoni propulsion (also known by the German term *Entspannungsschwimmen*).

Ecology

Insect ecology is the scientific study of how insects, individually or as a community, interact with the surrounding environment or ecosystem. Insects play one of the most important roles in their ecosystems, which includes many roles, such as soil turning and aeration, dung burial, pest control, pollination and wildlife nutrition. An example is the beetles, which are scavengers that feed on dead animals and fallen trees and thereby recycle biological materials into forms found useful by other organisms. These insects, and others, are responsible for much of the process by which topsoil is created.

Defense and Predation

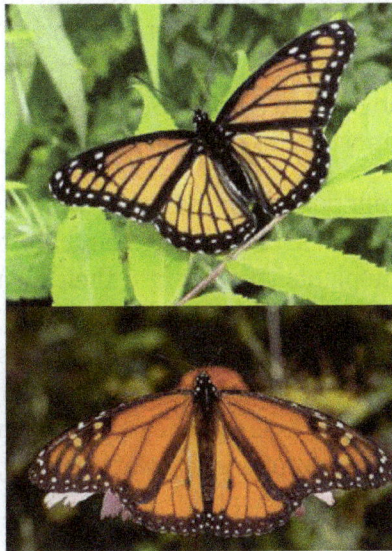

Perhaps one of the most well-known examples of mimicry,
the viceroy butterfly (top) appears very similar to the noxious-tasting monarch butterfly (bottom).

Insects are mostly soft bodied, fragile and almost defenseless compared to other, larger lifeforms. The immature stages are small, move slowly or are immobile, and so all stages are exposed to predation and parasitism. Insects then have a variety of defense strategies to avoid being attacked by predators or parasitoids. These include camouflage, mimicry, toxicity and active defense.

Camouflage is an important defense strategy, which involves the use of coloration or shape to blend into the surrounding environment. This sort of protective coloration is common and widespread among beetle families, especially those that feed on wood or vegetation, such as many of the leaf beetles (family Chrysomelidae) or weevils. In some of these species, sculpturing or various colored scales or hairs cause the beetle to resemble bird dung or other inedible objects. Many of those that live in sandy environments blend in with the coloration of the substrate. Most phasmids are known for effectively replicating the forms of sticks and leaves, and the bodies of some species (such as *O. macklotti* and *Palophus centaurus*) are covered in mossy or lichenous outgrowths that supplement their disguise. Some species have the ability to change color as their surroundings shift (*B. scabrinota*, *T. californica*). In a further behavioral adaptation to supplement crypsis, a number of species have been noted to perform a rocking motion where the body is swayed from side to side that is thought to reflect the movement of leaves or twigs swaying in the breeze. Another method by which stick insects avoid predation and resemble twigs is by feigning death (catalepsy), where the insect enters a motionless state that can be maintained for a long period. The nocturnal feeding habits of adults also aids Phasmatodea in remaining concealed from predators.

Another defense that often uses color or shape to deceive potential enemies is mimicry. A number of longhorn beetles (family Cerambycidae) bear a striking resemblance to wasps, which helps them avoid predation even though the beetles are in fact harmless. Batesian and Müllerian mimicry complexes are commonly found in Lepidoptera. Genetic polymorphism and natural selection give rise to otherwise edible species (the mimic) gaining a survival advantage by resembling inedible species (the model). Such a mimicry complex is referred to as *Batesian* and is most commonly known by the mimicry by the limenitidine viceroy butterfly of the inedible danaine monarch. Later research has discovered that the viceroy is, in fact more toxic than the monarch and this resemblance should be considered as a case of Müllerian mimicry. In Müllerian mimicry, inedible species, usually within a taxonomic order, find it advantageous to resemble each other so as to reduce the sampling rate by predators who need to learn about the insects' inedibility. Taxa from the toxic genus *Heliconius* form one of the most well known Müllerian complexes.

Chemical defense is another important defense found amongst species of Coleoptera and Lepidoptera, usually being advertised by bright colors, such as the monarch butterfly. They obtain their toxicity by sequestering the chemicals from the plants they eat into their own tissues. Some Lepidoptera manufacture their own toxins. Predators that eat poisonous butterflies and moths may become sick and vomit violently, learning not to eat those types of species; this is actually the basis of Müllerian mimicry. A predator who has previously eaten a poisonous lepidopteran may avoid other species with similar markings in the future, thus saving many other species as well. Some ground beetles of the Carabidae family can spray chemicals from their abdomen with great accuracy, to repel predators.

Pollination

European honey bee carrying pollen in a pollen basket back to the hive

Pollination is the process by which pollen is transferred in the reproduction of plants, thereby enabling fertilisation and sexual reproduction. Most flowering plants require an animal to do the transportation. While other animals are included as pollinators, the majority of pollination is done by insects. Because insects usually receive benefit for the pollination in the form of energy rich nectar it is a grand example of mutualism. The various flower traits (and combinations thereof) that differentially attract one type of pollinator or another are known as pollination syndromes. These arose through complex plant-animal adaptations. Pollinators find flowers through bright colorations, including ultraviolet, and attractant pheromones. The study of pollination by insects is known as *anthecology*.

Parasitism

Many insects are parasites of other insects such as the parasitoid wasps. These insects are known as entomophagous parasites. They can be beneficial due to their devastation of pests that can destroy crops and other resources. Many insects have a parasitic relationship with humans such as the mosquito. These insects are known to spread diseases such as malaria and yellow fever and because of such, mosquitoes indirectly cause more deaths of humans than any other animal.

Relationship to Humans

As Pests

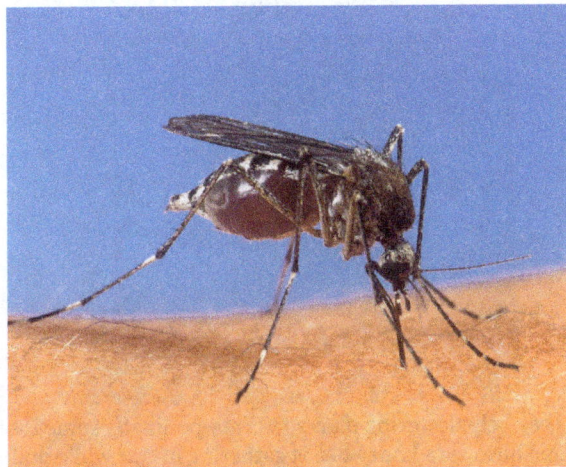

Aedes aegypti, a parasite, is the vector of dengue fever and yellow fever.

Many insects are considered pests by humans. Insects commonly regarded as pests include those that are parasitic (*e.g.* lice, bed bugs), transmit diseases (mosquitoes, flies), damage structures (termites), or destroy agricultural goods (locusts, weevils). Many entomologists are involved in various forms of pest control, as in research for companies to produce insecticides, but increasingly rely on methods of biological pest control, or biocontrol. Biocontrol uses one organism to reduce the population density of another organism — the pest — and is considered a key element of integrated pest management.

Despite the large amount of effort focused at controlling insects, human attempts to kill pests with insecticides can backfire. If used carelessly, the poison can kill all kinds of organisms in the area, including insects' natural predators, such as birds, mice and other insectivores. The effects of DDT's use exemplifies how some insecticides can threaten wildlife beyond intended populations of pest insects.

In Beneficial Roles

Although pest insects attract the most attention, many insects are beneficial to the environment and to humans. Some insects, like wasps, bees, butterflies and ants, pollinate flowering plants.

Pollination is a mutualistic relationship between plants and insects. As insects gather nectar from different plants of the same species, they also spread pollen from plants on which they have previously fed. This greatly increases plants' ability to cross-pollinate, which maintains and possibly even improves their evolutionary fitness. This ultimately affects humans since ensuring healthy crops is critical to agriculture. As well as pollination ants help with seed distribution of plants. This helps to spread the plants which increases plant diversity. This leads to an overall better environment. A serious environmental problem is the decline of populations of pollinator insects, and a number of species of insects are now cultured primarily for pollination management in order to have sufficient pollinators in the field, orchard or greenhouse at bloom time. Another solution, as shown in Delaware, has been to raise native plants to help support native pollinators like *L. vierecki*. Insects also produce useful substances such as honey, wax, lacquer and silk. Honey bees have been cultured by humans for thousands of years for honey, although contracting for crop pollination is becoming more significant for beekeepers. The silkworm has greatly affected human history, as silk-driven trade established relationships between China and the rest of the world.

Because they help flowering plants to cross-pollinate, some insects are
critical to agriculture. This European honey bee is gathering nectar while pollen collects on its body.

A robberfly with its prey, a hoverfly. Insectivorous relationships
such as these help control insect populations.

Insectivorous insects, or insects which feed on other insects, are beneficial to humans because they eat insects that could cause damage to agriculture and human structures. For example, aphids feed on crops and cause problems for farmers, but ladybugs feed on aphids, and can be used as a means to get significantly reduce pest aphid populations. While birds are perhaps more visible predators of insects, insects themselves account for the vast majority of insect consumption. Ants also help control animal populations by consuming small vertebrates. Without predators to keep them in check, insects can undergo almost unstoppable population explosions.

Insects are also used in medicine, for example fly larvae (maggots) were formerly used to treat wounds to prevent or stop gangrene, as they would only consume dead flesh. This treatment is finding modern usage in some hospitals. Recently insects have also gained attention as potential sources of drugs and other medicinal substances. Adult insects, such as crickets and insect larvae of various kinds, are also commonly used as fishing bait.

In Research

Insects play important roles in biological research. For example, because of its small size, short generation time and high fecundity, the common fruit fly *Drosophila melanogaster* is a model organism for studies in the genetics of higher eukaryotes. *D. melanogaster* has been an essential part of studies into principles like genetic linkage, interactions between genes, chromosomal genetics, development, behavior and evolution. Because genetic systems are well conserved among eukaryotes, understanding basic cellular processes like DNA replication or transcription in fruit flies can help to understand those processes in other eukaryotes, including humans. The genome of *D. melanogaster* was sequenced in 2000, reflecting the organism's important role in biological research. It was found that 70% of the fly genome is similar to the human genome, supporting the evolution theory.

The common fruitfly *Drosophila melanogaster* is one of the most widely used organisms in biological research.

As Food

In some cultures, insects, especially deep-fried cicadas, are considered to be delicacies, whereas in other places they form part of the normal diet. Insects have a high protein content for their mass, and some authors suggest their potential as a major source of protein in human nutrition. In most first-world countries, however, entomophagy (the eating of insects), is taboo. Since it is impossible to entirely eliminate pest insects from the human food chain, insects are inadvertently present in many foods, especially grains. Food safety laws in many countries do not prohibit insect parts in food, but rather limit their quantity. According to cultural materialist anthropologist Marvin Harris, the eating of insects is taboo in cultures that have other protein sources such as fish or livestock.

Due to the abundance of insects and a worldwide concern of food shortages, the Food and Agriculture Organisation of the United Nations considers that the world may have to, in the future, regard the prospects of eating insects as a food staple. Insects are noted for their nutrients, having a high content of protein, minerals and fats and are eaten by one-third of the global population.

In Culture

Scarab beetles held religious and cultural symbolism in Old Egypt, Greece and some shamanistic Old World cultures. The ancient Chinese regarded cicadas as symbols of rebirth or immortality. In Mesopotamian literature, the epic poem of Gilgamesh has allusions to Odonata which signify the impossibility of immortality. Amongst the Aborigines of Australia of the Arrernte language groups, honey ants and witchety grubs served as personal clan totems. In the case of the 'San' bush-men of the Kalahari, it is the praying mantis which holds much cultural significance including creation and zen-like patience in waiting.

Crustacean

Crustaceans form a large, diverse arthropod taxon which includes such familiar animals as crabs, lobsters, crayfish, shrimp, krill, woodlice and barnacles. The crustacean group is usually treated as a subphylum, and thanks to recent molecular studies it is now well accepted that the crustacean group is paraphyletic, and comprises all animals in the Pancrustacea clade other than hexapods. In other words, some crustaceans are more closely related to insects and other hexapods than they are to certain other crustaceans.

The 67,000 described species range in size from *Stygotantulus stocki* at 0.1 mm (0.004 in), to the Japanese spider crab with a leg span of up to 3.8 m (12.5 ft) and a mass of 20 kg (44 lb). Like other arthropods, crustaceans have an exoskeleton, which they moult to grow. They are distinguished from other groups of arthropods, such as insects, myriapods and chelicerates, by the possession of biramous (two-parted) limbs, and by their larval forms, such as the nauplius stage of branchiopods and copepods.

Most crustaceans are free-living aquatic animals, but some are terrestrial (e.g. woodlice), some are parasitic (e.g. Rhizocephala, fish lice, tongue worms) and some are sessile (e.g. barnacles). The group has an extensive fossil record, reaching back to the Cambrian, and includes living fossils such as *Triops cancriformis*, which has existed apparently unchanged since the Triassic period. More than 10 million tons of crustaceans are produced by fishery or farming for human consumption, the majority of it being shrimp and prawns. Krill and copepods are not as widely fished, but may be the animals with the greatest biomass on the planet, and form a vital part of the food chain. The scientific study of crustaceans is known as carcinology (alternatively, *malacostracology*, *crustaceology* or *crustalogy*), and a scientist who works in carcinology is a carcinologist.

Structure

The body of a crustacean is composed of segments, which are grouped into three regions: the *cephalon* or head, the thorax, and the *pleon* or abdomen. The head and thorax may be fused together to form a cephalothorax, which may be covered by a single large carapace. The crustacean body is protected by the hard exoskeleton, which must be moulted for the animal to grow. The shell around each somite can be divided into a dorsal tergum, ventral sternum and a lateral pleuron. Various parts of the exoskeleton may be fused together.

A shed carapace of a lady crab, part of the hard exoskeleton

Each somite, or body segment can bear a pair of appendages: on the segments of the head, these include two pairs of antennae, the mandibles and maxillae; the thoracic segments bear legs, which may be specialised as pereiopods (walking legs) and maxillipeds (feeding legs). The abdomen bears pleopods, and ends in a telson, which bears the anus, and is often flanked by uropods to form a tail fan. The number and variety of appendages in different crustaceans may be partly responsible for the group's success.

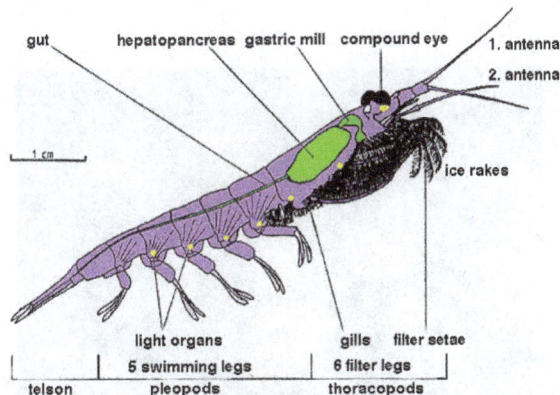

Body structure of a typical crustacean – krill

Crustacean appendages are typically biramous, meaning they are divided into two parts; this includes the second pair of antennae, but not the first, which is usually uniramous, the exception being in the Class Malacostraca where the antennules may be generally biramous or even triramous. It is unclear whether the biramous condition is a derived state which evolved in crustaceans, or whether the second branch of the limb has been lost in all other groups. Trilobites, for instance, also possessed biramous appendages.

The main body cavity is an open circulatory system, where blood is pumped into the haemocoel by a heart located near the dorsum. Malacostraca have haemocyanin as the oxygen-carrying pigment, while copepods, ostracods, barnacles and branchiopods have haemoglobins. The alimentary canal consists of a straight tube that often has a gizzard-like "gastric mill" for grinding food and a pair of digestive glands that absorb food; this structure goes in a spiral format. Structures that function as kidneys are located near the antennae. A brain exists in the form of ganglia close to the antennae, and a collection of major ganglia is found below the gut.

In many decapods, the first (and sometimes the second) pair of pleopods are specialised in the male for sperm transfer. Many terrestrial crustaceans (such as the Christmas Island red crab) mate seasonally and return to the sea to release the eggs. Others, such as woodlice, lay their eggs on land, albeit in damp conditions. In most decapods, the females retain the eggs until they hatch into free-swimming larvae.

Ecology

The majority of crustaceans are aquatic, living in either marine or freshwater environments, but a few groups have adapted to life on land, such as terrestrial crabs, terrestrial hermit crabs, and woodlice. Marine crustaceans are as ubiquitous in the oceans as insects are on land. The majority of crustaceans are also motile, moving about independently, although a few taxonomic units are parasitic and live attached to their hosts (including sea lice, fish lice, whale lice, tongue worms, and *Cymothoa exigua*, all of which may be referred to as "crustacean lice"), and adult barnacles live a sessile life – they are attached headfirst to the substrate and cannot move independently. Some branchiurans are able to withstand rapid changes of salinity and will also switch hosts from marine to non-marine species. Krill are the bottom layer and the most important part of the food chain in Antarctic animal communities. Some crustaceans are significant invasive species, such as the Chinese mitten crab, *Eriocheir sinensis*, and the Asian shore crab, *Hemigrapsus sanguineus*.

Life Cycle

Eggs of *Potamon fluviatile*, a freshwater crab

Zoea larva of the European lobster, *Homarus gammarus*

Mating System

The majority of crustaceans have separate sexes, and reproduce sexually. A small number are hermaphrodites, including barnacles, remipedes, and Cephalocarida. Some may even change sex during the course of their life. Parthenogenesis is also widespread among crustaceans, where viable eggs are produced by a female without needing fertilisation by a male. This occurs in many branchiopods, some ostracods, some isopods, and certain "higher" crustaceans, such as the *Marmorkrebs* crayfish.

Eggs

In many groups of crustaceans, the fertilised eggs are simply released into the water column, while others have developed a number of mechanisms for holding on to the eggs until they are ready to hatch. Most decapods carry the eggs attached to the pleopods, while peracarids, notostracans, anostracans, and many isopods form a brood pouch from the carapace and thoracic limbs. Female Branchiura do not carry eggs in external ovisacs but attach them in rows to rocks and other objects. Most leptostracans and krill carry the eggs between their thoracic limbs; some copepods carry their eggs in special thin-walled sacs, while others have them attached together in long, tangled strings.

Larvae

Crustaceans exhibit a number of larval forms, of which the earliest and most characteristic is the nauplius. This has three pairs of appendages, all emerging from the young animal's head, and a single naupliar eye. In most groups, there are further larval stages, including the zoea (pl. zoeæ or zoeas). This name was given to it when naturalists believed it to be a separate species. It follows the nauplius stage and precedes the post-larva. Zoea larvae swim with their thoracic appendages, as opposed to nauplii, which use cephalic appendages, and megalopa, which use abdominal appendages for swimming. It often has spikes on its carapace, which may assist these small organisms in maintaining directional swimming. In many decapods, due to their accelerated development, the zoea is the first larval stage. In some cases, the zoea stage is followed by the mysis stage, and in others, by the megalopa stage, depending on the crustacean group involved.

Classification

The name "crustacean" dates from the earliest works to describe the animals, including those of Pierre Belon and Guillaume Rondelet, but the name was not used by some later authors, including Carl Linnaeus, who included crustaceans among the "Aptera" in his *Systema Naturae*. The earliest nomenclaturally valid work to use the name "Crustacea" was Morten Thrane Brünnich's *Zoologiæ Fundamenta* in 1772, although he also included chelicerates in the group.

The subphylum Crustacea comprises almost 67,000 described species, which is thought to be just $\frac{1}{10}$ to $\frac{1}{100}$ of the total number as the majority of species remain as yet undiscovered. Although most crustaceans are small, their morphology varies greatly and includes both the largest arthropod in the world – the Japanese spider crab with a leg span of 3.7 metres (12 ft) – and the smallest, the 1-micrometre-long (0.00004 in) *Stygotantulus stocki*. Despite their diversity of form, crustaceans are united by the special larval form known as the nauplius.

The exact relationships of the Crustacea to other taxa are not completely settled as of April 2012.

Studies based on morphology led to the Pancrustacea hypothesis, in which Crustacea and Hexapoda (insects and allies) are sister groups. More recent studies using DNA sequences suggest that Crustacea is paraphyletic, with the hexapods nested within a larger Pancrustacea clade.

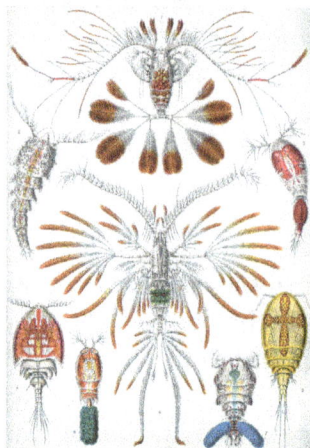

Copepods, from Ernst Haeckel's 1904 work *Kunstformen der Natur*

Although the classification of crustaceans has been quite variable, the system used by Martin and Davis largely supersedes earlier works. Mystacocarida and Branchiura, here treated as part of Maxillopoda, are sometimes treated as their own classes. Six classes are usually recognised:

Decapods, from Ernst Haeckel's 1904 work *Kunstformen der Natur*

Class	Members	Orders	Photo
Branchiopoda	brine shrimp fairy shrimp water fleas tadpole shrimp clam shrimp	Anostraca Lipostraca Notostraca Laevicaudata Spinicaudata Cyclestherida Cladocera	 *Daphnia pulex* (Cladocera)

Remipedia		Nectiopoda	*Speleonectes tanumekes* (Speleonectidae)
Maxillopoda	barnacles copepods	Calanoida Pedunculata Sessilia *c.* 20 others	*Chthamalus stellatus* (Sessilia)
Ostracoda	seed shrimp	Myodocopida Halocyprida Platycopida Podocopida	Cylindroleberididae
Malacostraca	crabs lobsters crayfish shrimp krill mantis shrimp woodlice hooded shrimp scuds sandhoppers *etc.*	Decapoda Isopoda Amphipoda Stomatopoda *c.* 12 others	*Gammarus roeseli* (Amphipoda)

Fossil Record

Eryma mandelslohi, a fossil decapod from the Jurassic of Bissingen an der Teck, Germany.

Crustaceans have a rich and extensive fossil record, which begins with animals such as *Canadaspis* and *Perspicaris* from the Middle Cambrian age Burgess Shale. Most of the major groups of crustaceans appear in the fossil record before the end of the Cambrian, namely the Branchiopoda, Maxillopoda (including barnacles and tongue worms) and Malacostraca; there is some debate as to whether or not Cambrian animals assigned to Ostracoda are truly ostracods, which would otherwise start in the Ordovician. The only classes to appear later are the Cephalocarida, which have no fossil record, and the Remipedia, which were first described from the fossil *Tesnusocaris goldichi*, but do not appear until the Carboniferous. Most of the early crustaceans are rare, but fossil crustaceans become abundant from the Carboniferous period onwards.

Norway lobsters on sale at a Spanish market

Within the Malacostraca, no fossils are known for krill, while both Hoplocarida and Phyllopoda contain important groups that are now extinct as well as extant members (Hoplocarida: mantis shrimp are extant, while Aeschronectida are extinct; Phyllopoda: Canadaspidida are extinct, while Leptostraca are extant). Cumacea and Isopoda are both known from the Carboniferous, as are the first true mantis shrimp. In the Decapoda, prawns and polychelids appear in the Triassic, and shrimp and crabs appear in the Jurassic; however, the great radiation of crustaceans occurred in the Cretaceous, particularly in crabs, and may have been driven by the adaptive radiation of their main predators, bony fish. The first true lobsters also appear in the Cretaceous.

Consumption by Humans

Many crustaceans are consumed by humans, and nearly 10,700,000 tons were produced in 2007; the vast majority of this output is of decapod crustaceans: crabs, lobsters, shrimp, crawfish, and prawns. Over 60% by weight of all crustaceans caught for consumption are shrimp and prawns, and nearly 80% is produced in Asia, with China alone producing nearly half the world's total. Non-decapod crustaceans are not widely consumed, with only 118,000 tons of krill being caught, despite krill having one of the greatest biomasses on the planet.

Coral

Corals are marine invertebrates in the class Anthozoa of phylum Cnidaria. They typically live in compact colonies of many identical individual polyps. The group includes the important reef builders that inhabit tropical oceans and secrete calcium carbonate to form a hard skeleton.

A coral "group" is a colony of myriad genetically identical polyps. Each polyp is a sac-like animal typically only a few millimeters in diameter and a few centimeters in length. A set of tentacles surround a central mouth opening. An exoskeleton is excreted near the base. Over many generations, the colony thus creates a large skeleton that is characteristic of the species. Individual heads grow by asexual reproduction of polyps. Corals also breed sexually by spawning: polyps of the same species release gametes simultaneously over a period of one to several nights around a full moon.

Although some corals can catch small fish and plankton using stinging cells on their tentacles, most corals obtain the majority of their energy and nutrients from photosynthetic unicellular dinoflagellates in the genus *Symbiodinium* that live within their tissues. These are commonly known as zooxanthellae and the corals that contain them are zooxanthellate corals. Such corals require sunlight and grow in clear, shallow water, typically at depths shallower than 60 metres (200 ft). Corals are major contributors to the physical structure of the coral reefs that develop in tropical and subtropical waters, such as the enormous Great Barrier Reef off the coast of Queensland, Australia.

Other corals do not rely on zooxanthellae and can live in much deeper water, with the cold-water genus *Lophelia* surviving as deep as 3,000 metres (9,800 ft). Some have been found on the Darwin Mounds, north-west of Cape Wrath, Scotland. Corals have also been found as far north as off the coast of Washington State and the Aleutian Islands.

Taxonomy

In his *Scala Naturae*, Aristotle classified corals as "zoophyta" ("plant-animals"), animals that had characteristics of plants and were therefore hypothetically in between animals and plants. The Persian polymath Al-Biruni (d. 1048) classified sponges and corals as animals, arguing that they respond to touch. Nevertheless, people believed corals to be plants until the eighteenth century, when William Herschel used a microscope to establish that coral had the characteristic thin cell membranes of an animal.

The phylogeny of Anthozoans is not clearly understood and a number of different models have been proposed. Within the Hexacorallia, the sea anemones, coral anemones and stony corals may constitute a monophyletic grouping united by their eight-fold symmetry and cnidocyte trait. The Octocorallia appears to be monophyletic, and primitive members of this group may have been stolonate. The cladogram presented here comes from a 2014 study by Stampar *et al.* which was based on the divergence of mitochondrial DNA within the group and on nuclear markers.

Corals are classified in the class Anthozoa of the phylum Cnidaria. They are divided into three subclasses, Hexacorallia, Octocorallia, and Ceriantharia. The Hexacorallia include the stony corals, the sea anemones and the zoanthids. These groups have polyps that generally have 6-fold symmetry. The Octocorallia include blue coral, soft corals, sea pens, and gorgonians (sea fans and sea whips). These groups have polyps with 8-fold symmetry, each polyp having eight tentacles and eight mesenteries. Ceriantharia are the tube-dwelling anemones.

Fire corals are not true corals, being in the order Anthomedusa (sometimes known as Anthoathecata) of the class Hydrozoa.

Anatomy

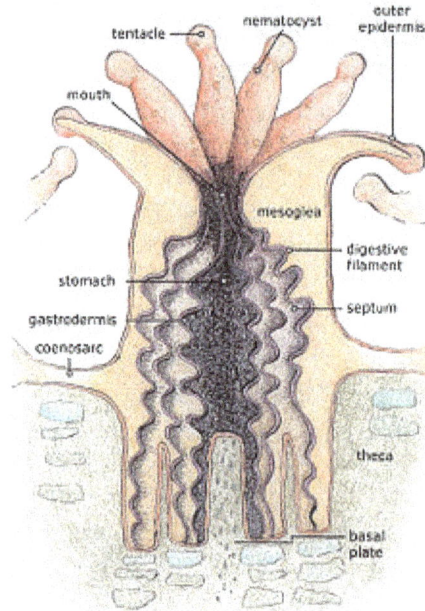

Anatomy of a stony coral polyp

Corals are sessile animals in the class Anthozoa and differ from most other cnidarians in not having a medusa stage in their life cycle. The body unit of the animal is a polyp. Most corals are colonial, the initial polyp budding to produce another and the colony gradually developing from this small start. In stony corals, also known as hard corals, the polyps produce a skeleton composed of calcium carbonate to strengthen and protect the organism. This is deposited by the polyps and by the coenosarc, the living tissue that connects them. The polyps sit in cup-shaped depressions in the skeleton known as corallites. Colonies of stony coral are very variable in appearance; a single species may adopt an encrusting, plate-like, bushy, columnar or massive solid structure, the various forms often being linked to different types of habitat, with variations in light level and water movement being significant.

In soft corals, there is no stony skeleton but the tissues are often toughened by the presence of tiny skeletal elements known as sclerites, which are made from calcium carbonate. Soft corals are very variable in form and most are colonial. A few soft corals are stolonate, but the polyps of most are connected by sheets of coenosarc. In some species this is thick and the polyps are deeply embedded. Some soft corals are encrusting or form lobes. Others are tree-like or whip-like and have a central axial skeleton embedded in the tissue matrix. This is composed either of a fibrous protein called gorgonin or of a calcified material. In both stony and soft corals, the polyps can be retracted, with stony corals relying on their hard skeleton and cnidocytes for defence against predators, with soft corals generally relying on chemical defences in the form of toxic substances present in the tissues known as terpenoids.

The polyps of stony corals have six-fold symmetry while those of soft corals have eight. The mouth of each polyp is surrounded by a ring of tentacles. In stony corals these are cylindrical and taper to a point, but in soft corals they are pinnate with side branches known as pinnules. In some tropical species these are reduced to mere stubs and in some they are fused to give a paddle-like appearance. In most corals, the tentacles are retracted by day and spread out at night to catch plankton and other small organisms. Shallow water species of both stony and soft corals can be zooxanthellate, the corals supplementing their plankton diet with the products of photosynthesis produced by these symbionts. The polyps interconnect by a complex and well-developed system of gastrovascular canals, allowing significant sharing of nutrients and symbionts.

Montastraea cavernosa polyps with tentacles extended

Ecology

Feeding

Polyps feed on a variety of small organisms, from microscopic zooplankton to small fish. The polyp's tentacles immobilize or kill prey using their nematocysts. These cells carry venom which they rapidly release in response to contact with another organism. A dormant nematocyst discharges in response to nearby prey touching the trigger (cnidocil). A flap (operculum) opens and its stinging apparatus fires the barb into the prey. The venom is injected through the hollow filament to immobilise the prey; the tentacles then manoeuvre the prey to the mouth.

The tentacles then contract to bring the prey into the stomach. Once the prey is digested, the stomach reopens, allowing the elimination of waste products and the beginning of the next hunting cycle. They can scavenge drifting organic molecules and dissolved organic molecules.

Intracellular Symbionts

Many corals, as well as other cnidarian groups such as *Aiptasia* (a sea anemone) form a symbiotic relationship with a class of dinoflagellate algae, zooxanthellae of the genus *Symbiodinium*. *Aiptasia*, a familiar pest among coral reef aquarium hobbyists, serves as a valuable model organism in the study of cnidarian-algal symbiosis. Typically, each polyp harbors one species of algae. Via photosynthesis, these provide energy for the coral, and aid in calcification. As much as 30% of the tissue of a polyp may be algal material.

The algae benefit from a safe place to live and consume the polyp's carbon dioxide and nitrogenous waste. Due to the strain the algae can put on the polyp, stress on the coral often drives them to eject the algae. Mass ejections are known as coral bleaching, because the algae contribute to coral's brown coloration; other colors, however, are due to host coral pigments, such as green fluorescent proteins (GFPs). Ejection increases the polyp's chance of surviving short-term stress—they can regain algae, possibly of a different species at a later time. If the stressful conditions persist, the polyp eventually dies.

Reproduction

Corals can be both gonochoristic (unisexual) and hermaphroditic, each of which can reproduce sexually and asexually. Reproduction also allows coral to settle in new areas.

Sexual

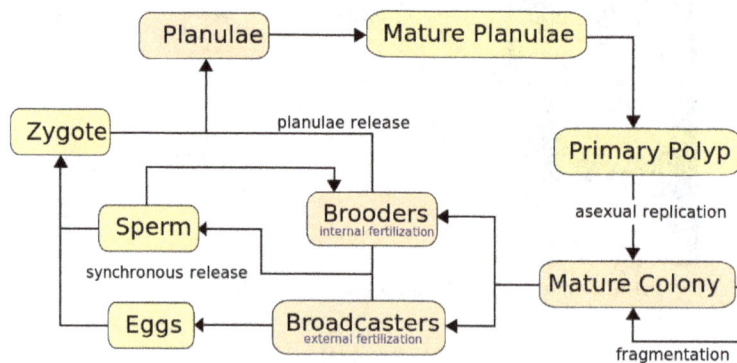

Life cycles of broadcasters and brooders

Corals predominantly reproduce sexually. About 25% of hermatypic corals (stony corals) form single sex (gonochoristic) colonies, while the rest are hermaphroditic.

Broadcasters

About 75% of all hermatypic corals "broadcast spawn" by releasing gametes—eggs and sperm—into the water to spread offspring. The gametes fuse during fertilization to form a microscopic larva called a planula, typically pink and elliptical in shape. A typical coral colony forms several thousand larvae per year to overcome the odds against formation of a new colony.

A male great star coral, *Montastraea cavernosa*, releasing sperm into the water.

Synchronous spawning is very typical on the coral reef, and often, even when multiple species are present, all corals spawn on the same night. This synchrony is essential so male and female gametes can meet. Corals rely on environmental cues, varying from species to species, to determine the proper time to release gametes into the water. The cues involve temperature change, lunar cycle, day length, and possibly chemical signalling. Synchronous spawning may form hybrids and is perhaps involved in coral speciation. The immediate cue is most often sunset, which cues the release. The spawning event can be visually dramatic, clouding the usually clear water with gametes.

Brooders

Brooding species are most often ahermatypic (not reef-building) in areas of high current or wave action. Brooders release only sperm, which is negatively buoyant, sinking on to the waiting egg carriers who harbor unfertilized eggs for weeks. Synchronous spawning events sometimes occurs even with these species. After fertilization, the corals release planula that are ready to settle.

Planulae

Planulae exhibit positive phototaxis, swimming towards light to reach surface waters, where they drift and grow before descending to seek a hard surface to which they can attach and begin a new colony. They also exhibit positive sonotaxis, moving towards sounds that emanate from the reef and away from open water. High failure rates afflict many stages of this process, and even though millions of gametes are released by each colony, few new colonies form. The time from spawning to settling is usually two to three days, but can be up to two months. The larva grows into a polyp and eventually becomes a coral head by asexual budding and growth.

Asexual

Basal plates (calices) of *Orbicella annularis* showing multiplication by budding (small central plate) and division (large double plate).

Budding involves splitting a smaller polyp from an adult. As the new polyp grows, it forms its body parts. The distance between the new and adult polyps grows, and with it, the coenosarc (the common body of the colony). Budding can be intratentacular, from its oral discs, producing same-sized polyps within the ring of tentacles, or extratentacular, from its base, producing a smaller polyp.

Within a coral head, the genetically identical polyps reproduce asexually, either by budding (gemmation) or by dividing, whether longitudinally or transversely.

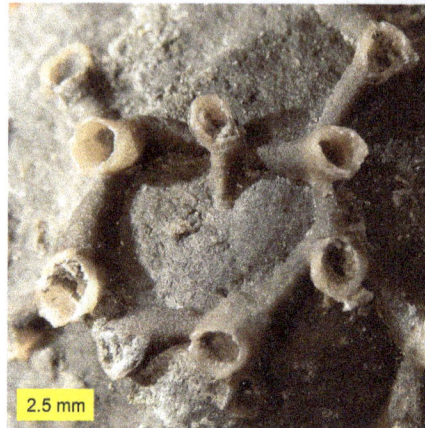

Tabulate coral *Aulopora* (Devonian) showing initial budding

Division forms two polyps that each become as large as the original. Longitudinal division begins when a polyp broadens and then divides its coelenteron (body), effectively splitting along its length. The mouth divides and new tentacles form. The two polyps thus created then generate their missing body parts and exoskeleton. Transversal division occurs when polyps and the exoskeleton divide transversally into two parts. This means one has the basal disc (bottom) and the other has the oral disc (top); the new polyps must separately generate the missing pieces.

Asexual reproduction offers the benefits of high reproductive rate, delaying senescence, and replacement of dead modules, as well as geographical distribution.

Colony Division

Whole colonies can reproduce asexually, forming two colonies with the same genotype. The possible mechanisms include fission, bailout and fragmentation. Fission occurs in some corals, especially among the family Fungiidae, where the colony splits into two or more colonies during early developmental stages. Bailout occurs when a single polyp abandons the colony and settles on a different substrate to create a new colony. Fragmentation involves individuals broken from the colony during storms or other disruptions. The separated individuals can start new colonies.

Reefs

Locations of coral reefs around the world

Many corals in the order Scleractinia are hermatypic, meaning that they are involved in building reefs. Most such corals obtain some of their energy from zooxanthellae in the genus *Symbiod-*

inium. These are symbiotic photosynthetic dinoflagellates which require sunlight; reef-forming corals are therefore found mainly in shallow water. They secrete calcium carbonate to form hard skeletons that become the framework of the reef. However, not all reef-building corals in shallow water contain zooxanthellae, and some deep water species, living at depths to which light cannot penetrate, form reefs but do not harbour the symbionts.

Staghorn coral (*Acropora cervicornis*) is an important hermatypic coral from the Caribbean.

There are various types of shallow-water coral reef, including fringing reefs, barrier reefs and atolls; most occur in tropical and subtropical seas. They are very slow-growing, adding perhaps one centimetre (0.4 in) in height each year. The Great Barrier Reef is thought to have been laid down about two million years ago. Over time, corals fragment and die, sand and rubble accumulates between the corals, and the shells of clams and other molluscs decay to form a gradually evolving calcium carbonate structure. Coral reefs are extremely diverse marine ecosystems hosting over 4,000 species of fish, massive numbers of cnidarians, molluscs, crustaceans, and many other animals.

Evolutionary History

2.0 cm

Solitary rugose coral (*Grewingkia*) in three views; Ordovician, southeastern Indiana.

Although corals first appeared in the Cambrian period, some 542 million years ago, fossils are extremely rare until the Ordovician period, 100 million years later, when rugose and tabulate corals became widespread. Paleozoic corals often contained numerous endobiotic symbionts.

Tabulate corals occur in limestones and calcareous shales of the Ordovician and Silurian periods, and often form low cushions or branching masses of calcite alongside rugose corals. Their numbers began to decline during the middle of the Silurian period, and they became extinct at the end of the Permian period, 250 million years ago.

Rugose or horn corals became dominant by the middle of the Silurian period, and became extinct early in the Triassic period. The rugose corals existed in solitary and colonial forms, and were also composed of calcite.

The scleractinian corals filled the niche vacated by the extinct rugose and tabulate species. Their fossils may be found in small numbers in rocks from the Triassic period, and became common in the Jurassic and later periods. Scleractinian skeletons are composed of a form of calcium carbonate known as aragonite. Although they are geologically younger than the tabulate and rugose corals, the aragonite of their skeletons is less readily preserved, and their fossil record is accordingly less complete.

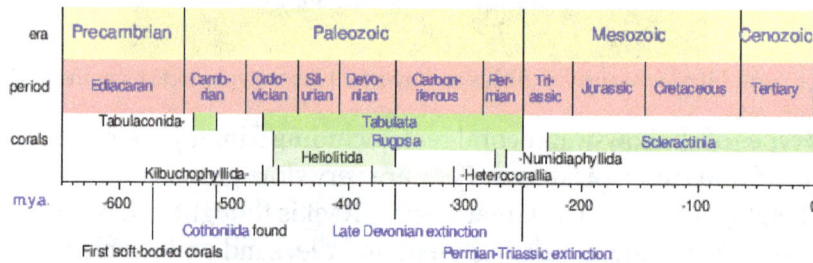

Timeline of the major coral fossil record and developments from 650 m.y.a. to present.

At certain times in the geological past, corals were very abundant. Like modern corals, these ancestors built reefs, some of which ended as great structures in sedimentary rocks. Fossils of fellow reef-dwellers algae, sponges, and the remains of many echinoids, brachiopods, bivalves, gastropods, and trilobites appear along with coral fossils. This makes some corals useful index fossils. Coral fossils are not restricted to reef remnants, and many solitary fossils may be found elsewhere, such as *Cyclocyathus*, which occurs in England's Gault clay formation.

Threats

Coral reefs are under stress around the world. In particular, coral mining, agricultural and urban runoff, pollution (organic and inorganic), overfishing, blast fishing, disease, and the digging of canals and access into islands and bays are localized threats to coral ecosystems. Broader threats are sea temperature rise, sea level rise and pH changes from ocean acidification, all associated with greenhouse gas emissions. In 1998, 16% of the world's reefs died as a result of increased water temperature.

Approximately 10% of the world's coral reefs are dead. About 60% of the world's reefs are at risk due to human-related activities. The threat to reef health is particularly strong in Southeast Asia, where 80% of reefs are endangered. Over 50% of the world's coral reefs may be destroyed by 2030; as a result, most nations protect them through environmental laws.

In the Caribbean and tropical Pacific, direct contact between ~40–70% of common seaweeds and coral causes bleaching and death to the coral via transfer of lipid-soluble metabolites. Seaweed and algae proliferate given adequate nutrients and limited grazing by herbivores such as parrotfish.

A healthy coral reef has a striking level of biodiversity in many forms of marine life.

Water temperature changes of more than 1–2 °C (1.8–3.6 °F) or salinity changes can kill some species of coral. Under such environmental stresses, corals expel their Symbiodinium; without them coral tissues reveal the white of their skeletons, an event known as coral bleaching.

Submarine springs found along the coast of Mexico's Yucatán Peninsula produce water with a naturally low pH (relatively high acidity) providing conditions similar to those expected to become widespread as the oceans absorb carbon dioxide. Surveys discovered multiple species of live coral that appeared to tolerate the acidity. The colonies were small and patchily distributed, and had not formed structurally complex reefs such as those that compose the nearby Mesoamerican Barrier Reef System.

Protection

Marine Protected Areas (MPAs), Biosphere reserves, marine parks, national monuments world heritage status, fishery management and habitat protection can protect reefs from anthropogenic damage.

Many governments now prohibit removal of coral from reefs, and inform coastal residents about reef protection and ecology. While local action such as habitat restoration and herbivore protection can reduce local damage, the longer-term threats of acidification, temperature change and sea-level rise remain a challenge.

To eliminate destruction of corals in their indigenous regions, projects have been started to grow corals in non-tropical countries.

Relation to Humans

Local economies near major coral reefs benefit from an abundance of fish and other marine creatures as a food source. Reefs also provide recreational scuba diving and snorkeling tourism. These activities can damage coral but international projects such as Green Fins that encourage dive and snorkel centres to follow a Code of Conduct have been proven to mitigate these risks.

Live coral is highly sought after for aquaria. Soft corals are easier to maintain in captivity than hard corals.

Jewelry

6-strand necklace, Navajo (Native American), ca. 1920s, Brooklyn Museum.

Corals' many colors give it appeal for necklaces and other jewelry. Intensely red coral is prized as a gemstone. Sometimes called fire coral, it is not the same as fire coral. Red coral is very rare because of overharvesting.

Always considered a precious mineral, "the Chinese have long associated red coral with auspiciousness and longevity because of its color and its resemblance to deer antlers (so by association, virtue, long life, and high rank". It reached its height of popularity during the Manchu or Qing Dynasty (1644-1911) when it was almost exclusively reserved for the emperor's use either in the form of coral beads (often combined with pearls) for court jewelry or as decorative Penjing (decorative miniature mineral trees). Coral was known as *shanhu* in Chinese. The "early-modern 'coral network' [began in] the Mediterranean Sea [and found its way] to Qing China via the English East India Company". There were strict rules regarding its use in a code established by the Qianlong Emperor in 1759.

Medicine

Depiction of coral in the Juliana Anicia Codex, a copy, written in
Constantinople in 515 AD, of Dioscorides' 1st century AD Greek work.
The facing page states that coral can be used to treat ulcers.

In medicine, chemical compounds from corals are used to treat cancer, AIDS and pain, and for other uses. Coral skeletons, e.g. *Isididae* are also used for bone grafting in humans. Coral Calx, known as Praval Bhasma in Sanskrit, is widely used in traditional system of Indian medicine as a supplement in the treatment of a variety of bone metabolic disorders associated with calcium deficiency. In classical times ingestion of pulverized coral, which consists mainly of the weak base calcium carbonate, was recommended for calming stomach ulcers by Galen and Dioscorides.

Construction

Tabulate coral (a syringoporid); Boone limestone (Lower Carboniferous)
near Hiwasse, Arkansas, scale bar is 2.0 cm.

Coral reefs in places such as the East African coast are used as a source of building material. Ancient (fossil) coral limestone, notably including the Coral Rag Formation of the hills around Oxford (England), was once used as a building stone, and can be seen in some of the oldest buildings in that city including the Saxon tower of St Michael at the Northgate, St. George's Tower of Oxford Castle, and the mediaeval walls of the city.

Climate Research

Annual growth bands in some corals, such as the deep sea bamboo corals (*Isididae*), may be among the first signs of the effects of ocean acidification on marine life. The growth rings allow geologists to construct year-by-year chronologies, a form of incremental dating, which underlie high-resolution records of past climatic and environmental changes using geochemical techniques.

Certain species form communities called microatolls, which are colonies whose top is dead and mostly above the water line, but whose perimeter is mostly submerged and alive. Average tide level limits their height. By analyzing the various growth morphologies, microatolls offer a low resolution record of sea level change. Fossilized microatolls can also be dated using Radiocarbon dating. Such methods can help to reconstruct Holocene sea levels.

Increasing sea temperatures in tropical regions (~1 degree C) the last century have caused major coral bleaching, death, and therefore shrinking coral populations since although they are able to adapt and acclimate, it is uncertain if this evolutionary process will happen quickly enough to prevent major reduction of their numbers.

Though coral have large sexually-reproducing populations, their evolution can be slowed by abundant asexual reproduction. Gene flow is variable among coral species. According to the biogeography of coral species gene flow cannot be counted on as a dependable source of adaptation as they are very stationary organisms. Also, coral longevity might factor into their adaptivity.

However, adaptation to climate changes has been demonstrated in many cases. These are usually due to a shift in coral and zooxanthellae genotypes. These shifts in allelic frequencies have progressed toward more tolerant types of zooxanthellae. Scientists found that a certain scleractinian zooxanthella is becoming more common where sea temperature is high. Symbionts able to tolerate warmer water seem to photosynthesise more slowly, implying an evolutionary trade-off.

In the Gulf of Mexico, where sea temperatures are rising, cold-sensitive staghorn and elkhorn coral have shifted in location. Not only have the symbionts and specific species been shown to shift, but there seems to be a certain growth rate favorable to selection. Slower-growing but more heat-tolerant corals have become more common. The changes in temperature and acclimation are complex. Some reefs in current shadows represent a refugium location that will help them adjust to the disparity in the environment even if eventually the temperatures may rise more quickly there than in other locations. This separation of populations by climatic barriers causes a realized niche to shrink greatly in comparison to the old fundamental niche.

Geochemistry

Corals are shallow, colonial organisms that integrate $\delta^{18}O$ and trace elements into their skeletal aragonite (polymorph of calcite) crystalline structures, as they grow. Geochemistry anomalies within the crystalline structures of corals represent functions of temperature, salinity and oxygen isotopic composition. Such geochemical analysis can help with climate modeling.

Strontium/Calcium Ratio Anomaly

Time can be attributed to coral geochemistry anomalies by correlating strontium/calcium minimums with sea surface temperature (SST) maximums to data collected from NINO 3.4 SSTA.

Oxygen Isotope Anomaly

The comparison of coral strontium/calcium minimums with sea surface temperature maximums, data recorded from NINO 3.4 SSTA, time can be correlated to coral strontium/calcium and $\delta^{18}O$ variations. To confirm accuracy of the annual relationship between Sr/Ca and $\delta^{18}O$ variations, a perceptible association to annual coral growth rings confirms the age conversion. Geochronology is established by the blending of Sr/Ca data, growth rings, and stable isotope data. El Nino-Southern Oscillation (ENSO) is directly related to climate fluctuations that influence coral $\delta^{18}O$ ratio from local salinity variations associated with the position of the South Pacific convergence zone (SPCZ) and can be used for ENSO modeling.

Sea Surface Temperature and Sea Surface Salinity

The global moisture budget is primarily being influenced by tropical sea surface temperatures from the position of the Intertropical Convergence Zone (ITCZ). The Southern Hemisphere has a unique meteorological feature positioned in the southwestern Pacific Basin called the South Pacific Convergence Zone (SPCZ), which contains a perennial position within the Southern Hemisphere. During ENSO warm periods, the SPCZ reverses orientation extending from the equator down south through Solomon Islands, Vanuatu, Fiji and towards the French Polynesian Islands; and due east towards South America affecting geochemistry of corals in tropical regions.

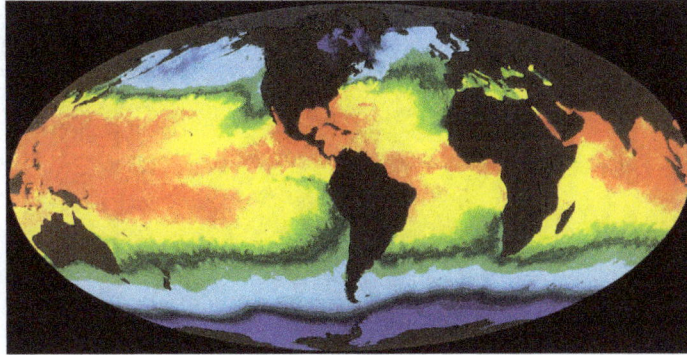

Global sea surface temperature (SST)

Geochemical analysis of skeletal coral can be linked to sea surface salinity (SSS) and sea surface temperature (SST), from El Nino 3.4 SSTA data, of tropical oceans to seawater $\delta^{18}O$ ratio anomalies from corals. ENSO phenomenon can be related to variations in sea surface salinity (SSS) and sea surface temperature (SST) that can help model tropical climate activities.

Limited Climate Research on Current Species

Genus: *Porites lutea*

Climate research on live coral species is limited to a few studied species. Studying *Porites* coral provides a stable foundation for geochemical interpretations that is much simpler to physically extract data in comparison to *Platygyra* species where the complexity of *Platygyra* species skeletal structure creates difficulty when physically sampled, which happens to be one of the only multidecadal living coral records used for coral paleoclimate modeling.

Aquaria

The saltwater fishkeeping hobby has increasingly expanded, over recent years, to include reef tanks, fish tanks that include large amounts of live rock on which coral is allowed to grow and spread. These tanks are either kept in a natural-like state, with algae (sometimes in the form of an algae scrubber) and a deep sand bed providing filtration, or as "show tanks", with the rock kept largely bare of the algae and microfauna that would normally populate it, in order to appear neat and clean.

This dragon-eye zoanthid is a popular source of color in reef tanks

The most popular kind of coral kept is soft coral, especially zoanthids and mushroom corals, which are especially easy to grow and propagate in a wide variety of conditions, because they originate in enclosed parts of reefs where water conditions vary and lighting may be less reliable and direct. More serious fishkeepers may keep small polyp stony coral, which is from open, brightly lit reef conditions and therefore much more demanding, while large polyp stony coral is a sort of compromise between the two.

Aquaculture

Coral aquaculture, also known as *coral farming* or *coral gardening*, is the cultivation of corals for commercial purposes or coral reef restoration. Aquaculture is showing promise as a potentially effective tool for restoring coral reefs, which have been declining around the world. The process bypasses the early growth stages of corals when they are most at risk of dying. Coral fragments known as "seeds" are grown in nurseries then replanted on the reef. Coral is farmed by coral farmers who live locally to the reefs and farm for reef conservation or for income. It is also farmed by scientists for research, by businesses for the supply of the live and ornamental coral trade and by private aquarium hobbyists.

Arachnid

Arachnids are a class (Arachnida) of joint-legged invertebrate animals (arthropods), in the subphylum Chelicerata. All arachnids have eight legs, although the front pair of legs in some species has converted to a sensory function, while in other species, different appendages can grow large enough to take on the appearance of extra pairs of legs. The term is derived from the Greek word (*aráchnē*), meaning "spider". Spiders are the largest order in the class, which also includes scorpions, ticks, mites, harvestmen, and solifuges.

Almost all extant arachnids are terrestrial, living mainly on land. However, some inhabit freshwater environments and, with the exception of the pelagic zone, marine environments as well. They comprise over 100,000 named species.

Morphology

Almost all adult arachnids have eight legs, and arachnids may be easily distinguished from insects by this fact, since insects have six legs. However, arachnids also have two further pairs of appendages that have become adapted for feeding, defense, and sensory perception. The first pair, the chelicerae, serve in feeding and defense. The next pair of appendages, the pedipalps, have been adapted for feeding, locomotion, and/or reproductive functions. In Solifugae, the palps are quite leg-like, so that these animals appear to have ten legs. The larvae of mites and Ricinulei have only six legs; a fourth pair usually appears when they moult into nymphs. However, mites are variable: as well as eight, there are adult mites with six or even four legs.

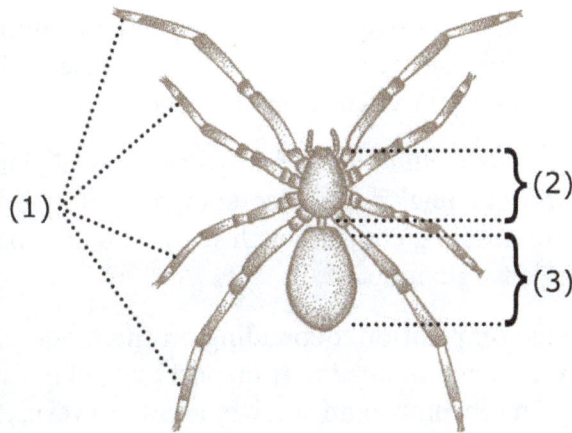

Basic characteristics of arachnids include four pairs of legs (1)
and a body divided into two tagmata: the cephalothorax (2) and the abdomen (3)

Arachnids are further distinguished from insects by the fact they do not have antennae or wings. Their body is organized into two tagmata, called the prosoma, or cephalothorax, and the opisthosoma, or abdomen. The cephalothorax is derived from the fusion of the cephalon (head) and the thorax, and is usually covered by a single, unsegmented carapace. The abdomen is segmented in the more primitive forms, but varying degrees of fusion between the segments occur in many groups. It is typically divided into a preabdomen and postabdomen, although this is only clearly visible in scorpions, and in some orders, such as the Acari, the abdominal sections are completely fused. A telson is present in scorpions, where it has been modified to a stinger, and in the Schizomida, whip scorpions and Palpigradi.

Like all arthropods, arachnids have an exoskeleton, and they also have an internal structure of cartilage-like tissue, called the endosternite, to which certain muscle groups are attached. The endosternite is even calcified in some Opiliones.

Locomotion

Most arachnids lack extensor muscles in the distal joints of their appendages. Spiders and whip-scorpions extend their limbs hydraulically using the pressure of their hemolymph. Solifuges and some harvestmen extend their knees by the use of highly elastic thickenings in the joint cuticle. Scorpions, pseudoscorpions and some harvestmen have evolved muscles that extend two leg joints (the femur-patella and patella-tibia joints) at once. The equivalent joints of the pedipalps of scorpions though, are extended by elastic recoil.

Physiology

There are characteristics that are particularly important for the terrestrial lifestyle of arachnids, such as internal respiratory surfaces in the form of tracheae, or modification of the book gill into a book lung, an internal series of vascular lamellae used for gas exchange with the air. While the tracheae are often individual systems of tubes, similar to those in insects, ricinuleids, pseudoscorpions, and some spiders possess sieve tracheae, in which several tubes arise in a bundle from a small chamber connected to the spiracle. This type of tracheal system has almost certainly evolved from the book lungs, and indicates that the tracheae of arachnids are not homologous with those of insects.

Further adaptations to terrestrial life are appendages modified for more efficient locomotion on land, internal fertilisation, special sensory organs, and water conservation enhanced by efficient excretory structures as well as a waxy layer covering the cuticle.

The excretory glands of arachnids include up to four pairs of coxal glands along the side of the prosoma, and one or two pairs of Malpighian tubules, emptying into the gut. Many arachnids have only one or the other type of excretory gland, although several do have both. The primary nitrogenous waste product in arachnids is guanine.

Arachnid blood is variable in composition, depending on the mode of respiration. Arachnids with an efficient tracheal system do not need to transport oxygen in the blood, and may have a reduced circulatory system. In scorpions and some spiders, however, the blood contains haemocyanin, a copper-based pigment with a similar function to haemoglobin in vertebrates. The heart is located in the forward part of the abdomen, and may or may not be segmented. Some mites have no heart at all.

Diet and Digestive System

Arachnids are mostly carnivorous, feeding on the pre-digested bodies of insects and other small animals. Only in the harvestmen and among mites, such as the house dust mite, is there ingestion of solid food particles, and thus exposure to internal parasites, although it is not unusual for spiders to eat their own silk. Several groups secrete venom from specialized glands to kill prey or enemies. Several mites and ticks are parasites, some of which are carriers of disease.

Arachnids produce digestive juices in their stomachs, and use their pedipalps and chelicerae to pour them over their dead prey. The digestive juices rapidly turn the prey into a broth of nutrients, which the arachnid sucks into a pre-buccal cavity located immediately in front of the mouth. Behind the mouth is a muscular, sclerotised pharynx, which acts as a pump, sucking the food through the mouth and on into the oesophagus and stomach. In some arachnids, the oesophagus also acts as an additional pump.

The stomach is tubular in shape, with multiple diverticula extending throughout the body. The stomach and its diverticula both produce digestive enzymes and absorb nutrients from the food. It extends through most of the body, and connects to a short sclerotised intestine and anus in the hind part of the abdomen.

Senses

Arachnids have two kinds of eyes: the lateral and median ocelli. The lateral ocelli evolved from compound eyes and may have a tapetum, which enhances the ability to collect light. With the exception of scorpions, which can have up to five pairs of lateral ocelli, there are never more than three pairs present. The median ocelli develop from a transverse fold of the ectoderm. The ancestors of modern arachnids probably had both types, but modern ones often lack one type or the other. The cornea of the eye also acts as a lens, and is continuous with the cuticle of the body. Beneath this is a transparent vitreous body, and then the retina and, if present, the tapetum. In most arachnids, the retina probably does not have enough light sensitive cells to allow the eyes to form a proper image.

In addition to the eyes, almost all arachnids have two other types of sensory organs. The most important to most arachnids are the fine sensory hairs that cover the body and give the animal its sense of touch. These can be relatively simple, but many arachnids also possess more complex structures, called trichobothria.

Finally, slit sense organs are slit-like pits covered with a thin membrane. Inside the pit, a small hair touches the underside of the membrane, and detects its motion. Slit sense organs are believed to be involved in proprioception, and possibly also hearing.

Reproduction

Arachnids may have one or two gonads, which are located in the abdomen. The genital opening is usually located on the underside of the second abdominal segment. In most species, the male transfers sperm to the female in a package, or spermatophore. Complex courtship rituals have evolved in many arachnids to ensure the safe delivery of the sperm to the female.

Arachnids usually lay yolky eggs, which hatch into immatures that resemble adults. Scorpions, however, are either ovoviviparous or viviparous, depending on species, and bear live young. In most arachnids only the females provide parental care, with harvestmen being one of the few exceptions.

Systematics

Phylogeny

The phylogenetic relationships among the main subdivisions of arthropods have been the subject of considerable research and dispute for many years. A consensus emerged from about 2010 onwards, based on both morphological and molecular evidence. Extant (living) arthropods are a monophyletic group and are divided into three main clades: chelicerates (including arachnids), pancrustaceans (the paraphyletic crustaceans plus insects and their allies), and myriapods (centipedes, millipedes and allies). The three groups are related as shown in the cladogram below. Including fossil taxa does not fundamentally alter this view, although it introduces some additional basal groups.

The extant chelicerates comprise two marine groups: sea spiders and horseshoe crabs, and the terrestrial arachnids. These are related as shown below. (Pycnogonida (sea spiders) may be excluded from the chelicerates, which are then identified as the group labelled "Euchelicerata".)

Discovering relationships within the arachnids has proved difficult as of March 2016, with successive studies producing different results. A study in 2014, based on the largest set of molecular data to date, concluded that there were systematic conflicts in the phylogenetic information, particularly affecting the orders Acariformes, Parasitiformes and Pseudoscorpiones, which have had much faster evolutionary rates. Analyses of the data using sets of genes with different evolutionary rates produced mutually incompatible phylogenetic trees. The authors favoured relationships shown by more slowly evolving genes, which demonstrated the monophyly of Chelicerata, Euchelicerata and Arachnida, as well as of some clades within the arachnids. The diagram summarizes their conclusions, based largely on the 200 most slowly evolving genes; dashed lines represent uncertain placements.

Arachnopulmonata

Tetrapulmonata, here consisting of Araneae, Amblypygi and Thelyphonida (Schizomida was not included in the study), received strong support. The addition of Scorpiones to produce a clade called Arachnopulmonata was also well supported. Pseudoscorpiones may also belong here, possibly as the sister of Scorpiones. Somewhat unexpectedly, there was support for a clade comprising Opiliones, Ricinulei and Solifugae, a combination not found in most other studies.

Subdivisions

The subdivisions of the arachnids are usually treated as orders. Historically, mites and ticks were treated as a single order, Acari. However, molecular phylogenetic studies suggest that the two groups do not form a single clade, with morphological similarities being due to convergence. They are now usually treated as two separate taxa – Acariformes, mites, and Parasitiformes, ticks – which may be ranked as orders or superorders. The arachnid subdivisions are listed below alphabetically; numbers of species are approximate.

- Acariformes – mites (32,000 species)

- Amblypygi – "blunt rump" tail-less whip scorpions with front legs modified into whip-like sensory structures as long as 25 cm or more (153 species)

- Araneae – spiders (40,000 species)

- †Haptopoda – extinct arachnids apparently part of the Tetrapulmonata, the group including spiders and whip scorpions (1 species)

- Opiliones – phalangids, harvestmen or daddy-long-legs (6,300 species)

- Palpigradi – microwhip scorpions (80 species)

- Parasitiformes – ticks (12,000 species)

- †Phalangiotarbi – extinct arachnids of uncertain affinity (30 species)

- Pseudoscorpionida – pseudoscorpions (3,000 species)

- Ricinulei – ricinuleids, hooded tickspiders (60 species)

- Schizomida – "split middle" whip scorpions with divided exoskeletons (220 species)

- Scorpiones – scorpions (2,000 species)

- Solifugae – solpugids, windscorpions, sun spiders or camel spiders (900 species)

- Thelyphonida (also called Uropygi) – whip scorpions or vinegaroons, forelegs modified into sensory appendages and a long tail on abdomen tip (100 species)

- †Trigonotarbida – extinct (late Silurian early Permian)

- †Uraraneida – extinct spider-like arachnids, but with a "tail" and no spinnerets (2 species)

It is estimated that 98,000 arachnid species have been described, and that there may be up to 600,000 in total.

Acari

Acari or Acarina is a taxon of arachnids that contains mites and ticks. Its fossil history goes back to the Devonian period, although there is also a questionable Ordovician record. The Devonian period was the time frame in which certain species of animals developed legs. In most modern treatments, the Acari is considered a subclass of Arachnida and is composed of two or three orders or superorders: Acariformes, Parasitiformes, and Opilioacariformes. Most acarines are minute to small (e.g. 0.080–1.00 mm), but the giants of the Acari (some ticks and red velvet mites) may reach lengths of 10–20 mm. It is estimated that over 50,000 species have been described (as of 1999) and that a million or more species are currently living. The study of mites and ticks is called acarology.

Only the faintest traces of primary segmentation remain in mites, the prosoma and opisthosoma being insensibly fused, and a region of flexible cuticle (the cirumcapitular furrow) separates the chelicerae and pedipalps from the rest of the body. This anterior body region is called the gnathosoma (or capitulum) and is also found in the Ricinulei. The remainder of the body is called the idiosoma and is unique to mites. Most adult mites have four pairs of legs, like other arachnids, but some have fewer. For example, gall mites like *Phyllocoptes variabilis* (superfamily Eriophyioidea) have a wormlike body with only two pairs of legs; some parasitic mites have only one or three pairs of legs in the adult stage. Larval and prelarval stages have a maximum of three pairs of legs; adult mites with only three pairs of legs may be called 'larviform'.

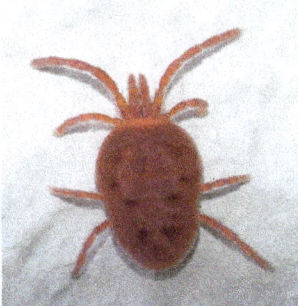

Trombidium holosericeum (a mite)

Ixodes hexagonus, a tick

Acarine ontogeny consists of an egg, a prelarval stage (often absent), a larval stage (hexapod except in Eriophyoidea, which have only two pairs of legs), and a series of nymphal stages. Larvae (and prelarvae) have a maximum of three pairs of legs (legs are often reduced to stubs or absent in prelarvae); the fourth pair of legs is added at the first nymphal stage.

Acarines live in practically every habitat, and include aquatic (freshwater and sea water) and terrestrial species. They outnumber other arthropods in the soil organic matter and detritus. Many are parasitic, and they affect both vertebrates and invertebrates. Most parasitic forms are external parasites, while the free living forms are generally predaceous and may even be used to control undesirable arthropods. Others are detritivores that help to break down forest litter and dead organic matter such as skin cells. Others still are plant feeders and may damage crops. Damage to crops is perhaps the most costly economic effect of mites, especially by the spider mites and their relatives (Tetranychoidea), earth mites (Penthaleidae), thread-footed mites (Tarsonemidae) and the gall and rust mites (Eriophyoidea). Some parasitic forms affect humans and other mammals, causing damage by their feeding, and can even be vectors of diseases such as scrub typhus and rickettsial pox. A well-known effect of mites on humans is their role as an allergen and the stimulation of asthma in people affected by the respiratory disease. The use of predatory mites (e.g. Phytoseiidae) in pest control and herbivorous mites that attack weeds is also important. An unquantified, but major positive contribution of the Acari is their normal functioning in ecosystems, especially their roles in the decomposer subsystem.

Amblypygi

Amblypygids are also known as tailless whip scorpions or cave spiders. Approximately 5 families, 17 genera and 136 species have been described. They are found in tropical and subtropical regions worldwide. Some species are subterranean; many are nocturnal. During the day, they may hide under logs, bark, stones, or leaves. They prefer a humid environment. Amblypygids may range from 5 to 40 mm. Their bodies are broad and highly flattened and the first pair of legs (the first walking legs in most arachnid orders) are modified to act as sensory organs. (Compare solifugids, uropygids, and schizomids.) These very thin modified legs can extend several times the length of body. They have no silk glands or venomous fangs, but can have prominent pincer-like pedipalps. Amblypygids often move about sideways on their six walking legs, with one "whip" pointed in the direction of travel while the other probes on either side of them. Prey are located with these "whips", captured with pedipalps, then torn to pieces with chelicerae. Fossilised amblypygids have been found dating back to the Carboniferous period.

An amblypygid

Amblypygids, particularly the species *Phrynus marginemaculatus* and *Damon diadema*, are thought to be one of the few species of arachnids that show signs of social behavior. Research conducted at Cornell University by entomologists suggests that a mother amblypygid comforts her young by gently caressing the offspring with her feelers. Further, when two or more siblings were placed in an unfamiliar environment, such as a cage, they would seek each other out and gather back in a group.

Araneae

Araneus diadematus

Araneae, or spiders, are the most familiar of the arachnids, and the most species-rich with around 40,000 described species. All spiders produce silk, a thin, strong protein strand extruded by the spider from spinnerets most commonly found on the end of the abdomen. Many species use it to trap insects in webs, although there are many species that hunt freely. Silk can be used to aid in climbing, form smooth walls for burrows, build egg sacs, wrap prey, temporarily hold sperm, and even fly, among other applications.

All spiders except those in the families Uloboridae and Holarchaeidae, and in the suborder Mesothelae (together about 350 species) can inject venom to protect themselves or to kill and liquefy prey. Only about 200 species, however, have bites that can pose health problems to humans. Many larger species' bites may be painful, but will not produce lasting health concerns.

Spiders are found all over the world, from the tropics to the Arctic, with some extreme species even living underwater in silken domes that they supply with air, and on the tops of the highest mountains.

Haptopoda

Haptopoda is an extinct order known exclusively from a few specimens from the Upper Carboniferous of the United Kingdom. It is monotypic, i.e. has only one species, *Plesiosiro madeleyi*. Relationships with other arachnids are obscure, but closest relatives may be the Amblypygi, Thelyphonida and Schizomida of the tetrapulmonate clade – a result which has been reflected in cladistic analyses.

Opiliones

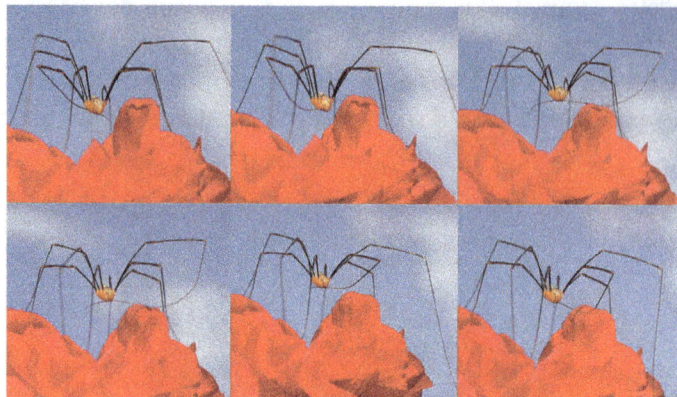

Male *Opilio canestrinii* cleaning its legs

Opiliones (formerly Phalangida, and better known as harvestmen or daddy longlegs) are arachnids that are harmless to people and are known for their exceptionally long walking legs, compared to their body size. As of December 2011, over 6,500 species of harvestmen have been discovered worldwide. The order Opiliones is divided into five suborders: Cyphophthalmi, Eupnoi, Dyspnoi, Laniatores, and the recently described Tetrophthalmi. Well-preserved fossils have been found in the 410-million year old Rhynie cherts of Scotland and 305-million-year-old rocks from France; they look surprisingly modern, suggesting that the basic structure of the harvestmen has not changed much since then.

The difference between harvestmen and spiders is that in harvestmen the two main body sections (the abdomen or opisthosoma with ten segments and the cephalothorax or prosoma) are nearly joined, so that they appear to be one oval structure. In more advanced species, the first five abdominal segments are often fused into a dorsal shield called the scutum, which is normally fused with the carapace. Sometimes this shield is only present in males. The two hindmost abdominal segments may be reduced or separated in the middle on the surface to form two plates lying next to each other. The second pair of legs is longer than the others and works as antennae. They have a single pair of eyes in the middle of their heads, oriented sideways. They have a pair of prosomatic scent glands that secrete a peculiar smelling fluid when disturbed. Harvestmen do not have spinnerets and do not possess poison glands, posing absolutely no danger to humans. They breathe through tracheae. Between the base of the fourth pair of legs and the abdomen is a pair of spiracles, one opening on each side. In more active species, spiracles are also found upon the tibia of the legs. They have a gonopore on the ventral cephalothorax, and copulation is direct, as the male has a penis (while the female has an ovipositor).

Typical body length does not exceed 7 millimetres (0.28 in) even in the largest species. However, leg span is much larger and can exceed 160 mm (6.3 in). Most species live for a year. Many species are omnivorous, eating primarily small insects and all kinds of plant material and fungi; some are scavengers of the decays of any dead animal, bird dung and other fecal material. They are mostly nocturnal and coloured in hues of brown, although there are a number of diurnal species that have vivid patterns in yellow, green and black with varied reddish and blackish mottling and reticulation.

Palpigradi

Palpigradi, commonly known as "microwhip scorpions", are tiny cousins of the uropygid, or whip scorpion, no more than 3 mm in length. They have a thin, pale, segmented carapace that terminates in a whip-like flagellum, made up of 15 segments. The carapace is divided into two plates between the third and fourth leg set. They have no eyes. Some species have three pairs of book lungs, while others have no respiratory organs at all. Approximately 80 species of Palpigradi have been described worldwide, in the families Eukoeneniidae and Prokoeneniidae, with a total of seven genera.

They are believed to be predators like their larger relatives, feeding on minuscule insects in their habitat. Their mating habits are unknown, except that they lay only a few relatively large eggs at a time. Microwhip scorpions need a damp environment to survive, and they always hide from light, so they are commonly found in the moist earth under buried stones and rocks. They can be found on every continent, except in Arctic and Antarctic regions.

Phalangiotarbi

Phalangiotarbi is an extinct arachnid order known exclusively from the Palaeozoic (Devonian to Permian) of Europe and North America. The affinities of phalangiotarbids are obscure, with most authors favouring affinities with Opiliones (harvestmen) and/or Acari (mites and ticks). Phalangiotarbi has been recently proposed to be sister group to (Palpigradi+Tetrapulmonata): the taxon Megoperculata sensu Shultz (1990).

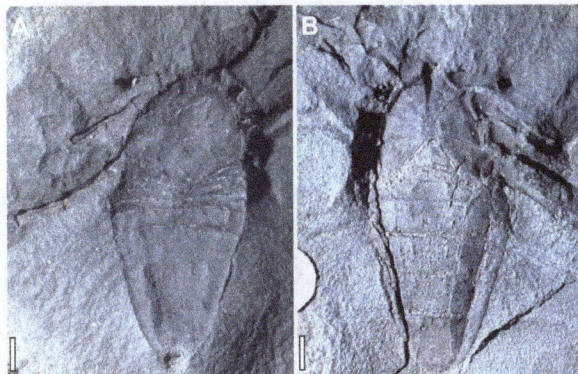

Fossil of *Goniotarbus angulatus* – dorsal (left) and ventral (right) surfaces

Pseudoscorpions

A pseudoscorpion on a printed page

Pseudoscorpions are small arthropods with a flat, pear-shaped body and pincers that resemble those of scorpions. They range from 2 to 8 mm (0.079 to 0.315 in) long. The opisthosoma is made up of twelve segments, each guarded by plate-like tergites above and sternites below. The abdomen is short and rounded at the rear, rather than extending into a segmented tail and stinger like true scorpions. The colour of the body can be yellowish-tan to dark-brown, with the paired claws often a contrasting colour. They may have two, four or no eyes. They have two very long pedipalps with palpal chelae (pincers) that strongly resemble the pincers found on a scorpion. The pedipalps generally consist of an immobile "hand" and "finger", with a separate movable finger controlled by an adductor muscle. A venom gland and duct are usually located in the mobile finger; the poison is used to capture and immobilise the pseudoscorpion's prey. During digestion, pseudoscorpions pour a mildly corrosive fluid over the prey, then ingest the liquefied remains. Pseudoscorpions spin silk from a gland in their jaws to make disk-shaped cocoons for mating, molting, or waiting out cold weather. Another trait they share with their closest relatives, the spiders, is breathing through spiracles. Most spiders have one pair of spiracles, and one of book lungs, but pseudoscorpions do not have book lungs.

There are more than 2,000 species of pseudoscorpions recorded. They range worldwide, even in temperate to cold regions, but have their most dense and diverse populations in the tropics and subtropics. The fossil record of pseudoscorpions dates back over 380 million years, to the Devonian period, near the time when the first land-animal fossils appear.

During the elaborate mating dance, the male of some pseudoscorpion species pulls a female over a spermatophore previously laid upon a surface. In other species, the male also pushes the sperm into the female genitals using the forelegs. The female carries the fertilised eggs in a brood pouch attached to her abdomen, and the young ride on the mother for a short time after they hatch. Up to two dozen young are hatched in a single brood; there may be more than one brood per year. The young go through three molts over the course of several years before reaching adulthood. Adult pseudoscorpions live 2 to 3 years. They are active in the warm months of the year, overwintering in silken cocoons when the weather grows cold.

Pseudoscorpions are generally beneficial to humans since they prey on clothes moth larvae, carpet beetle larvae, booklice, ants, mites, and small flies. They are small and inoffensive, and are rarely seen due to their size. They usually enter the home by "riding along" with larger insects (known as phoresy), or are brought in with firewood. They are often observed in bathrooms or laundry rooms, since they seek humidity. They may sometimes be found feeding on mites under the wing covers of certain beetles.

Ricinulei

Cryptocellus goodnighti

Ricinulei (hooded tickspiders) are 5–10 mm long. Their most notable feature is a "hood" that can be raised and lowered over the head; when lowered, it covers the mouth and the chelicerae. Ricinulei have no eyes. The pedipalps end in pincers that are small relative to their bodies, when compared to those of the related orders of scorpions and pseudoscorpions. The heavy-bodied abdomen forms a narrow pedicel, or waist, where it attaches to the prosoma. In males, the third pair of legs are modified to form copulatory organs. Malpighian tubules and a pair of coxal glands make up the excretory system. They have no lungs, as gas exchange takes place through the trachea.

Ricinulei are predators, feeding on other small arthropods. Little is known about their mating habits; the males have been observed using their modified third leg to transfer a spermatophore to the female. The eggs are carried under the mother's hood, until the young hatch into six-legged "larva", which later molt into their adult forms. Ricinulei require moisture to survive. Approximately 57 species of ricinuleids have been described worldwide, all in a single family that contains three genera.

Schizomida

Hubbardia pentapeltis female

Schizomida is an order of arachnids that tend to live in the top layer of soils. Schizomids present the prosoma covered by a large protopeltidium and smaller, paired, mesopeltidia and metapeltidia. There are no eyes. The opisthosoma is a smooth oval of 12 recognisable somites. The first is reduced and forms the pedicel. The last three are much constricted, forming the pygidium. The last somite bears the flagellum, which in this order is short and consists of not more than four segments.

The name means "split or cleaved middle", referring to the way the cephalothorax is divided into two separate plates. Like the Thelyphonida, Amblypygi, and Solifugae, the schizomids use only six legs for walking, having modified their first two legs to serve as sensory organs. They also have large well-developed pedipalps (pincers) just behind the sensory legs.

Scorpiones

Scorpio maurus palmatus

Scorpions (formally the order Scorpiones) are characterised by a metasoma (tail) comprising six segments, the last containing the scorpion's anus and bearing the telson (the sting). The telson, in turn, consists of the vesicle, which holds a pair of venom glands and the hypodermic aculeus, the venom-injecting barb. The abdomen's front half, the mesosoma, is made up of six segments. The first segment contains the sexual organs as well as a pair of vestigial and modified appendages forming a structure called the genital operculum. The second segment bears a pair of featherlike sensory organs known as the "pectines"; the final four segments each contain a pair of book lungs. The mesosoma is armored with chitinous plates, known as tergites on the upper surface and sternites on the lower surface.

The cuticle of scorpions is covered with hairs in some places that act like balance organs. An outer layer that makes them fluorescent green under ultraviolet light is called the hyaline layer. Newly molted scorpions do not glow until after their cuticle has hardened. The fluorescent hyaline layer can be intact in fossil rocks that are hundreds of millions of years old.

Scorpions are opportunistic predators of small arthropods and insects. They use their chelae (pincers) to catch the prey initially. Depending on the toxicity of their venom and size of their claws, they will then either crush the prey or inject it with neurotoxic venom. The neurotoxins consist of a variety of small proteins as well as sodium and potassium cations, which serve to interfere with neurotransmission in the victim. Scorpions use their venom to kill or paralyze their prey so that it can be eaten; in general, it is fast acting, allowing for effective prey capture. Scorpion venoms are optimised for action on other arthropods and therefore most scorpions are relatively harmless to humans; stings produce only local effects (such as pain, numbness or swelling). A few scorpion species, however, mostly in the family Buthidae, can be dangerous to humans. The scorpion that is responsible for the most human deaths is *Androctonus australis*, or fat-tailed scorpion of North Africa. The toxicity of *A. australis*'s venom is roughly half that of the deathstalker (*Leiurus quinquestriatus*), but since *A. australis* injects quite a bit more venom into its prey, it is the most deadly to humans. Human deaths normally occur in the young, elderly, or infirm; scorpions are generally unable to deliver enough venom to kill healthy adults. Some people, however, may be allergic to the venom of some species, in which case the scorpion's sting can more likely kill. A primary symptom of a scorpion sting is numbing at the injection site, sometimes lasting for several days. It has been found that scorpions have two types of venom: a translucent, weaker venom designed to stun only, and an opaque, more potent venom designed to kill heavier threats.

Unlike the majority of Arachnida species, scorpions are viviparous. The young are born one by one, and the brood is carried about on its mother's back until the young have undergone at least one moult. The young generally resemble their parents, requiring between five and seven moults to reach maturity. Scorpions have quite variable lifespans and the lifespan of most species is not known. The age range appears to be approximately 4–25 years (25 years being the maximum reported life span in the giant desert hairy scorpion, *Hadrurus arizonensis*). They are nocturnal and fossorial, finding shelter during the day in the relative cool of underground holes or undersides of rocks and coming out at night to hunt and feed. Scorpions prefer to live in areas where the temperature is 20–37 °C (68–99 °F), but may survive in the temperature range of 14–45 °C (57–113 °F).

Scorpions have been found in many fossil records, including coal deposits from the Carboniferous Period and in marine Silurian deposits. They are thought to have existed in some form since about 450 to 425 million years ago. They are believed to have an oceanic origin, with gills and a claw-like appendage that enabled them to hold onto rocky shores or seaweed.

Solifugae

Solifugae is a group of 900 species of arachnids, commonly known as camel spiders, wind scorpions, and sun spiders. The name derives from Latin, and means *those that flee from the sun*. Most Solifugae live in tropical or semitropical regions where they inhabit warm and arid habitats, but some species have been known to live in grassland or forest habitats. The most distinctive feature of Solifugae is their large chelicerae. Each of the two chelicerae are composed of two articles forming a powerful pincer; each article bears a variable number of teeth. Males in all families but Eremobatidae possess a flagellum on the basal article of the chelicera. Solifugae also have long pedipalps, which function as sense organs similar to insects' antennae and give the appearance of the two extra legs. Pedipalps terminate in reversible adhesive organs.

Solifugae are carnivorous or omnivorous, with most species feeding on termites, darkling beetles, and other small arthropods; however, solifugae have been videotaped consuming larger prey, such as lizards. Prey is located with the pedipalps and killed and cut into pieces by the chelicerae. The prey is then liquefied and the liquid ingested through the pharynx. Reproduction can involve direct or indirect sperm transfer; when indirect, the male emits a spermatophore on the ground and then inserts it with his chelicerae in the female's genital pore.

Galeodes sp.

Thelyphonida

A whip scorpion

The Thelyphonida (formerly Uropygi), commonly known as vinegarroons or whip scorpions, range from 25 to 85 mm in length; the largest species, of the genus *Mastigoproctus*, reaches 85 mm (3.3 in). Like the related orders Schizomida, Amblypygi, and Solifugae, the vinegarroons use only six legs for walking, having modified their first two legs to serve as antennae-like sensory organs. Many species also have very large scorpion-like pedipalps (pincers). They have one pair of eyes at the front of the cephalothorax and three on each side of the head. Whip scorpions have no poison glands, but they do have glands near the rear of their abdomen that can spray a combination of acetic acid and octanoic acid when they are bothered. Other species spray formic acid or chlorine. As of 2006, over 100 species have been described worldwide.

Whip scorpions are carnivorous, nocturnal hunters feeding mostly on insects but sometimes on worms and slugs. The prey is crushed between special teeth on the inside of the trochanters (the second segment of the leg) of the front legs. They are valuable in controlling cockroach and cricket populations.

Males secrete a sperm sac, which is transferred to the female. Up to 35 eggs are laid in a burrow, within a mucous membrane that preserves moisture. Mothers stay with the eggs and do not eat. The white young that hatch from the eggs climb onto their mother's back and attach themselves there with special suckers. After the first molt, they look like miniature whip scorpions, and leave the burrow; the mother dies soon after. The young grow slowly, going through three molts in about three years before reaching adulthood.

Vinegarroons are found in tropical and subtropical areas worldwide, usually in underground burrows that they dig with their pedipalps. They may also burrow under logs, rotting wood, rocks, and other natural debris. They enjoy humid, dark places and avoid the light.

Trigonotarbida

Fossil of *Kreischeria* species

The order Trigonotarbida is an extinct group of arachnids whose fossil record extends from the Silurian to the Lower Permian. They are known from several localities in North Asia, North America and Argentina. They superficially resemble spiders, to which they were clearly related - most cladistic analyses recover them in a clade with Thelyphonida, Schizomida, Amblypygi and Araneae.

These early arachnids seem to have been adapted to stalking prey on the ground. They have been found within the very structure of ground-dwelling plants, possibly where they hid to await their prey. Trigonotarbids are currently among the oldest known land arthropods. They lack silk glands on the opisthosoma and cheliceral poison glands, and most likely represented independent offshoots of the Arachnida.

Uraraneida

Uraraneida is a small extinct order of arachnids consisting of two fossil species found in the Devonian and the Permian. They were at one time identified as spiders, and are clearly closely related to them. However, they differ in a several respects. Silk is produced from spigots borne along the rear edges of plates on the underside of the abdomen, not on appendage-like spinnerets, as in spiders. They also have a long, jointed "tail" or flagellum at the end of the abdomen, after the anus, a feature lacking in spiders but present in some other arachnids, such as whip scorpions.

References

- Byrne, J.H., Castellucci, V.F. and Kandel,E.R., (1978). Contribution of individual mechanoreceptor sensory neurons to defensive gill-withdrawal reflex in Aplysia. Journal of Neurophysiology, 41: 418–431

- Brown, Lesley (1993). The New shorter Oxford English dictionary on historical principles. Oxford [Eng.]: Clarendon. ISBN 0-19-861271-0

- Todaro, Antonio. "Gastrotricha: Overview". Gastrotricha: World Portal. University of Modena & Reggio Emilia. Retrieved 2008-01-26

- Seeley T.D.; Visscher P.K.; Passino K.M. (2006). "Group decision making in honey bee swarms". American Scientist. 94: 220–229. doi:10.1511/2006.3.220

- Cees H. J. Hof (1998). "Fossil stomatopods (Crustacea: Malacostraca) and their phylogenetic impact". Journal of Natural History. 32 (10 & 11): 1567–1576. doi:10.1080/00222939800771101

- Wasserthal, Lutz T. (1998). Chapter 25: The Open Hemolymph System of Holometabola and Its Relation to the Tracheal Space. In "Microscopic Anatomy of Invertebrates". Wiley-Liss, Inc. ISBN 0-471-15955-7

- "Tree of Life Web Project. Version 1 January 1995 (temporary) of Arthropoda". Tree of Life Web Project. 1995. Retrieved 9 May 2009

- Engel, Michael S.; David A. Grimaldi (2004). "New light shed on the oldest insect". Nature. 427 (6975): 627–630. Bibcode:2004Natur.427..627E. PMID 14961119. doi:10.1038/nature02291

- Lukas, Roger; Lindstrom, Eric (1991). "The mixed layer of the western equatorial Pacific Ocean". Journal of Geophysical Research. 96 (S1): 3343–58. Bibcode:1991JGR....96.3343L. doi:10.1029/90JC01951

- Prewitt, Nancy L.; Larry S. Underwood; William Surver (2003). BioInquiry: making connections in biology. John Wiley. p. 289. ISBN 978-0-471-20228-8

- Mallet, Jim (12 June 2007). "Taxonomy of Lepidoptera: the scale of the problem". The Lepidoptera Taxome Project. University College, London. Retrieved 8 February 2011

- Russell Garwood; Gregory Edgecombe (2011). "Early terrestrial animals, evolution and uncertainty". Evolution, Education, and Outreach. 4 (3): 489–501. doi:10.1007/s12052-011-0357-y

- Diaz, James H. (2004). "The global epidemiology, syndromic classification, management, and prevention of spider bites". American Journal of Tropical Medicine and Hygiene. 71 (2): 239–250. PMID 15306718

- Schowalter, Timothy Duane (2006). Insect ecology: an ecosystem approach (2(illustrated) ed.). Academic Press. p. 572. ISBN 978-0-12-088772-9

- "Glossary of Lepidopteran and Odonate anatomy". Rare species atlas. Virginia Department of Conservation and Recreation. 2013. Retrieved 14 June 2013

- Schneiderman, Howard A. (1960). "Discontinuous respiration in insects: role of the spiracles". Biol. Bull. 119 (3): 494–528. JSTOR 1539265. doi:10.2307/1539265

- Hadley, Neil F. (1970). "Water relations of the desert scorpion Hadrurus arizonensis" (PDF). Journal of Experimental Biology. 53 (3): 547–558. PMID 5487163

- Lloyd, James E. (1984). "Occurrence of Aggressive Mimicry in Fireflies". The Florida Entomologist. 67 (3): 368–376. JSTOR 3494715. doi:10.2307/3494715

- Pierce, BA (2006). Genetics: A Conceptual Approach (2nd ed.). New York: W.H. Freeman and Company. p. 87. ISBN 0-7167-8881-0

- Brewer, Gary. "Social insects". North Dakota State University. Archived from the original on 21 March 2008. Retrieved 6 May 2009

- May, Robert M. (16 September 1988). "How Many Species Are There on Earth?". Science. 241 (4872): 1441–1449. JSTOR 1702670. PMID 17790039. doi:10.1126/science.241.4872.1441

3

An Integrated Study of Marine invertebrates

The category of invertebrates found in marine habitats is known as marine invertebrates. Echinoderm, cnidaria and tunicate are some of the examples of marine invertebrates. The aspects elucidated in this chapter are of vital importance, and provide a better understanding of marine invertebrates.

Marine Invertebrates

Ernst Haeckel's 96th plate, showing some marine invertebrates. Marine invertebrates have a large variety of body plans, which are currently categorised into over 30 phyla.

Marine invertebrates are the invertebrates that live in marine habitats. Invertebrate is a blanket term that includes all animals apart from the vertebrate members of the chordate phylum. Invertebrates lack a vertebral column, and some have evolved a shell or a hard exoskeleton. As on land and in the air, marine invertebrates have a large variety of body plans, and have been categorised into over 30 phyla. They make up most of the macroscopic life in the oceans.

Evolution

The earliest animals were marine invertebrates, that is, vertebrates came later. Animals are multicellular eukaryotes, and are distinguished from plants, algae, and fungi by lacking cell walls.

Marine invertebrates are animals that inhabit a marine environment apart from the vertebrate members of the chordate phylum; invertebrates lack a vertebral column. Some have evolved a shell or a hard exoskeleton.

Kimberella, an early mollusc important for understanding
the Cambrian explosion. Invertebrates are grouped into different phyla (body plans).

Opabinia an extinct, stem group arthropod that appeared in the Middle Cambrian

The earliest widely accepted animal fossils are the rather modern-looking cnidarians (the group that includes jellyfish, sea anemones and *Hydra*), possibly from around 580 Ma The Ediacara biota, which flourished for the last 40 million years before the start of the Cambrian, were the first animals more than a very few centimetres long. Many were flat and had a "quilted" appearance, and seemed so strange that there was a proposal to classify them as a separate kingdom, Vendozoa. Others, however, have been interpreted as early molluscs (*Kimberella*), echinoderms (*Arkarua*), and arthropods (*Spriggina, Parvancorina*). There is still debate about the classification of these specimens, mainly because the diagnostic features which allow taxonomists to classify more recent organisms, such as similarities to living organisms, are generally absent in the Ediacarans. However, there seems little doubt that *Kimberella* was at least a triploblastic bilaterian animal, in other words, an animal significantly more complex than the cnidarians.

The small shelly fauna are a very mixed collection of fossils found between the Late Ediacaran and Middle Cambrian periods. The earliest, *Cloudina*, shows signs of successful defense against predation and may indicate the start of an evolutionary arms race. Some tiny Early Cambrian shells almost certainly belonged to molluscs, while the owners of some "armor plates," *Halkieria* and *Microdictyon*, were eventually identified when more complete specimens were found in Cambrian lagerstätten that preserved soft-bodied animals.

In the 1970s there was already a debate about whether the emergence of the modern phyla was "explosive" or gradual but hidden by the shortage of Precambrian animal fossils. A re-analysis of fossils from the Burgess Shale lagerstätte increased interest in the issue when it revealed animals, such as *Opabinia*, which did not fit into any known phylum. At the time these were interpreted as evidence that the modern phyla had evolved very rapidly in the Cambrian explosion and that the Burgess Shale's "weird wonders" showed that the Early Cambrian was a uniquely experimental period of animal evolution. Later discoveries of similar animals and the development of new theoretical approaches led to the conclusion that many of the "weird wonders" were evolutionary "aunts" or "cousins" of modern groups—for example that *Opabinia* was a member of the lobopods, a group which includes the ancestors of the arthropods, and that it may have been closely related to the modern tardigrades. Nevertheless, there is still much debate about whether the Cambrian explosion was really explosive and, if so, how and why it happened and why it appears unique in the history of animals.

Classification

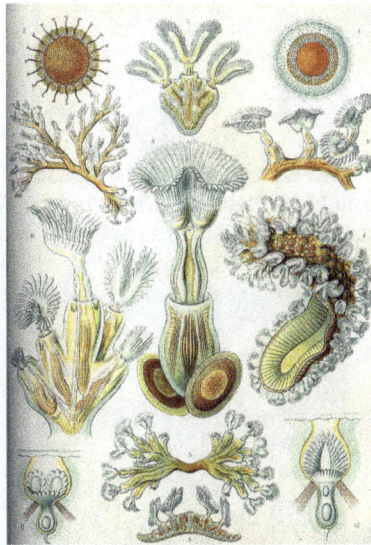

Bryozoa, from Ernst Haeckel's *Kunstformen der Natur*, 1904

Invertebrates are grouped into different phyla. Informally phyla can be thought of as a way of grouping organisms according to their body plan. A body plan refers to a blueprint which describes the shape or morphology of an organism, such as its symmetry, segmentation and the disposition of its appendages. The idea of body plans originated with vertebrates, which were grouped into one phylum. But the vertebrate body plan is only one of many, and invertebrates consist of many phyla or body plans. The history of the discovery of body plans can be seen as a movement from a worldview centred on vertebrates, to seeing the vertebrates as one body plan among many. Among the pioneering zoologists, Linnaeus identified two body plans outside the vertebrates; Cuvier identified three; and Haeckel had four, as well as the Protista with eight more, for a total of twelve. For comparison, the number of phyla recognised by modern zoologists has risen to 35.

Historically body plans were thought of as having evolved in rapidly during the Cambrian explosion, but a more nuanced understanding of animal evolution suggests a gradual development of body plans throughout the early Palaeozoic and beyond. More generally a phylum can be defined in two ways: as described above, as a group of organisms with a certain degree of morphological or developmental similarity (the phenetic definition), or a group of organisms with a certain degree of evolutionary relatedness (the phylogenetic definition).

As on land and in the air, invertebrates make up a great majority of all macroscopic life, as the vertebrates makes up a subphylum of one of over 30 known animal phyla, making the term almost meaningless for taxonomic purpose. Invertebrate sea life includes the following groups, some of which are phyla:

The 49th plate from Ernst Haeckel's *Kunstformen der Natur*, 1904, showing various sea anemones classified as Actiniae, in the Cnidaria phylum

"A variety of marine worms": plate from *Das Meer* by M.J. Schleiden (1804–1881)

- Acoela, among the most primitive bilateral animals;

- Annelida, (polychaetes and sea leeches);

- Brachiopoda, marine animals that have hard "valves" (shells) on the upper and lower surfaces ;

- Bryozoa, also known as moss animals or sea mats;

- Chaetognatha, commonly known as arrow worms, are a phylum of predatory marine worms that are a major component of plankton;

- Cephalochordata represented in the modern oceans by the lancelets (also known as Amphioxus);

- Cnidaria, such as jellyfish, sea anemones, and corals;

- Crustacea, including lobsters, crabs, shrimp, crayfish, barnacles, hermit crabs, mantis shrimps, and copepods;

- Ctenophora, also known as comb jellies, the largest animals that swim by means of cilia;

- Echinodermata, including sea stars, brittle stars, sea urchins, sand dollars, sea cucumbers, crinoids, and sea daisies;

- Echiura, also known as spoon worms;

- Gnathostomulids, slender to thread-like worms, with a transparent body that inhabit sand and mud beneath shallow coastal waters;

- Gastrotricha, often called hairy backs, found mostly interstitially in between sediment particles;

- Hemichordata, includes acorn worms, solitary worm-shaped organisms;

- Kamptozoa, goblet-shaped sessile aquatic animals, with relatively long stalks and a "crown" of solid tentacles, also called Entoprocta;

- Kinorhyncha, segmented, limbless animals, widespread in mud or sand at all depths, also called mud dragons;

- Loricifera, very small to microscopic marine sediment-dwelling animals only discovered in 1983;

- Merostomata; also known as horseshoe crabs;

- Mollusca, including shellfish, squid, octopus, whelks, *Nautilus*, cuttlefish, nudibranchs, scallops, sea snails, Aplacophora, Caudofoveata, Monoplacophora, Polyplacophora, and Scaphopoda;

- Myzostomida, a taxonomic group of small marine worms which are parasitic on crinoids or "sea lilies";

- Nemertinea, also known as "ribbon worms" or "proboscis worms";

- Orthonectida, a small phylum of poorly known parasites of marine invertebrates that are among the simplest of multi-cellular organisms;

- Phoronida, a phylum of marine animals that filter-feed with a lophophore (a "crown" of tentacles), and build upright tubes of chitin to support and protect their soft bodies;

- Placozoa, small, flattened, multicellular animals around 1 millimetre across and the simplest in structure. They have no regular outline, although the lower surface is somewhat concave, and the upper surface is always flattened;

- Porifera (sponges), multicellular organisms that have bodies full of pores and channels allowing water to circulate through them;

- Priapulida, or penis worms, are a phylum of marine worms that live marine mud. They are named for their extensible spiny proboscis, which, in some species, may have a shape like that of a human penis;

- Pycnogonida, also called sea spiders, are unrelated to spiders, or even to arachnids which they resemble;

- Sipunculida, also called peanut worms, is a group containing 144–320 species (estimates vary) of bilaterally symmetrical, unsegmented marine worms;

- Tunicata, also known as sea squirts or sea pork, are filter feeders attached to rocks or similarly suitable surfaces on the ocean floor;

- Some flatworms of the classes Turbellaria and Monogenea;

- Xenoturbella, a genus of bilaterian animals that contains only two marine worm-like species;

- Xiphosura, includes a large number of extinct lineages and only four recent species in the family Limulidae, which include the horseshoe crabs.

Arthropods total about 1,113,000 described extant species, molluscs about 85,000 and chordates about 52,000.

Marine Sponges

Sponges have no nervous, digestive or circulatory system

Sponges are animals of the phylum Porifera (Modern Latin for *bearing pores*). They are multicellular organisms that have bodies full of pores and channels allowing water to circulate through them, consisting of jelly-like mesohyl sandwiched between two thin layers of cells. They have unspecialized cells that can transform into other types and that often migrate between the main cell layers and the mesohyl in the process. Sponges do not have nervous, digestive or circulatory systems. Instead, most rely on maintaining a constant water flow through their bodies to obtain food and oxygen and to remove wastes.

Sponges are similar to other animals in that they are multicellular, heterotrophic, lack cell walls and produce sperm cells. Unlike other animals, they lack true tissues and organs, and have no body symmetry. The shapes of their bodies are adapted for maximal efficiency of water flow through the central cavity, where it deposits nutrients, and leaves through a hole called the osculum. Many sponges have internal skeletons of spongin and/or spicules of calcium carbonate or silicon dioxide. All sponges are sessile aquatic animals. Although there are freshwater species, the great majority are marine (salt water) species, ranging from tidal zones to depths exceeding 8,800 m (5.5 mi).

While most of the approximately 5,000–10,000 known species feed on bacteria and other food particles in the water, some host photosynthesizing micro-organisms as endosymbionts and these alliances often produce more food and oxygen than they consume. A few species of sponge that live in food-poor environments have become carnivores that prey mainly on small crustaceans.

| Stove-pipe sponge | Branching vase sponge | Venus' flower basket at a depth of 2572 meters | Barrel sponge |

Linnaeus mistakenly identified sponges as plants in the order Algae. For a long time thereafter sponges were assigned to a separate subkingdom, Parazoa (meaning *beside the animals*). They are now classified as a paraphyletic phylum from which the higher animals have evolved.

Marine Cnidarians

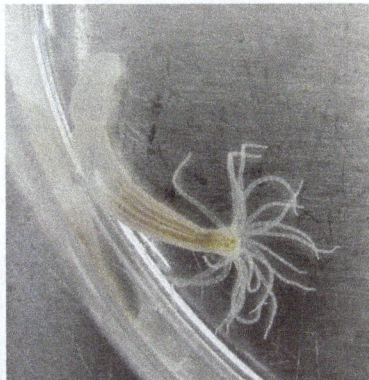

Cnidarians are the simplest animals with cells organised into tissues.
Yet the starlet sea anemone contains the same genes as those that form the vertebrate head.

Cnidarians (Greek for *nettle*) are distinguished by the presence of stinging cells, specialized cells that they use mainly for capturing prey. Cnidarians include corals, sea anemones, jellyfish and hydrozoans. They form a phylum containing over 10,000 species of animals found exclusively in aquatic (mainly marine) environments. Their bodies consist of mesoglea, a non-living jelly-like substance, sandwiched between two layers of epithelium that are mostly one cell thick. They have two basic body forms: swimming medusae and sessile polyps, both of which are radially symmetrical with mouths surrounded by tentacles that bear cnidocytes. Both forms have a single orifice and body cavity that are used for digestion and respiration.

Fossil cnidarians have been found in rocks formed about 580 million years ago. Fossils of cnidarians that do not build mineralized structures are rare. Scientists currently think cnidarians, ctenophores and bilaterians are more closely related to calcareous sponges than these are to other sponges, and that anthozoans are the evolutionary "aunts" or "sisters" of other cnidarians, and the most closely related to bilaterians.

Cnidarians are the simplest animals in which the cells are organised into tissues. The starlet sea anemone is used as a model organism in research. It is easy to care for in the laboratory and a protocol has been developed which can yield large numbers of embryos on a daily basis. There is a remarkable degree of similarity in the gene sequence conservation and complexity between the sea anemone and vertebrates. In particular, genes concerned in the formation of the head in vertebrates are also present in the anemone.

Sea anemones are common in tidepools

Portuguese man o' war

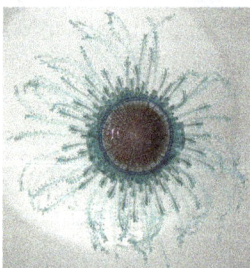

Porpita porpita

Turritopsis dohrnii, a small, biologically immortal jellyfis transfers its cells back to childhood.

Moon jellyfish, found in coastal waters around the world

Lion's mane jellyfish, largest known jellyfish

Marine Worms

Worms (Old English for *serpent*) typically have long cylindrical tube-like bodies and no limbs. Marine worms vary in size from microscopic to over 1 metre (3.3 ft) in length for some marine polychaete worms (bristle worms) and up to 58 metres (190 ft) for the marine nemertean worm (bootlace worm). Some marine worms occupy a small variety of parasitic niches, living inside the bodies of other animals, while others live more freely in the marine environment or by burrowing underground.

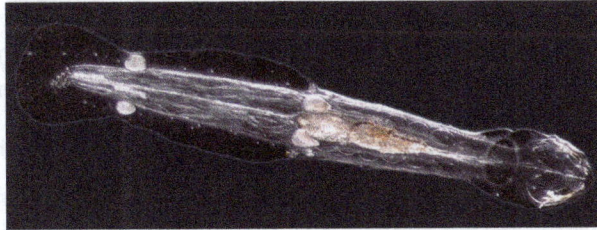

Arrow worms are predatory components of plankton worldwide.

Different groups of marine worms are related only distantly, so they are found in several different phyla such as the Annelida (segmented worms), Chaetognatha (arrow worms), Hemichordata, and Phoronida (horseshoe worms). Many of these worms have specialized tentacles used for exchanging oxygen and carbon dioxide and also may be used for reproduction. Some marine worms are tube worms, such as the giant tube worm which lives in waters near underwater volcanoes and can withstand temperatures up to 90 degrees Celsius.

Platyhelminthes (flatworms) form another worm phylum which includes a class Cestoda of parasitic tapeworms. The marine tapeworm *Polygonoporus giganticus*, found in the gut of sperm whales, can grow to over 30 m (100 ft).

Nematodes (roundworms) constitute a further worm phylum with tubular digestive systems and an opening at both ends. Over 25,000 nematode species have been described, of which more than half are parasitic. It has been estimated another million remain undescribed. They are ubiquitous in marine, freshwater and terrestrial environments, where they often outnumber other animals in both individual and species counts. They are found in every part of the earth's lithosphere, from the top of mountains to the bottom of oceanic trenches. By count they represent 90% of all animals on the ocean floor. Their numerical dominance, often exceeding a million individuals per square meter and accounting for about 80% of all individual animals on earth, their diversity of life cycles, and their presence at various trophic levels point at an important role in many ecosystems.

Echinoderms

Echinoderms (Greek for *spiny skin*) is a phylum which contains only marine invertebrates. The adults are recognizable by their radial symmetry (usually five-point) and include starfish, sea urchins, sand dollars, and sea cucumbers, as well as the sea lilies. Echinoderms are found at every ocean depth, from the intertidal zone to the abyssal zone. The phylum contains about 7000 living species, making it the second-largest grouping of deuterostomes (a superphylum), after the chordates (which include the vertebrates, such as birds, fishes, mammals, and reptiles).

Echinoderms are unique among animals in having bilateral symmetry at the larval stage, but five-fold symmetry (pentamerism, a special type of radial symmetry) as adults.

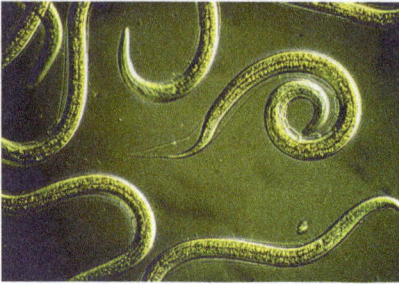

Nematodes are ubiquitous pseudocoelomates which can parasite marine plants and animals.

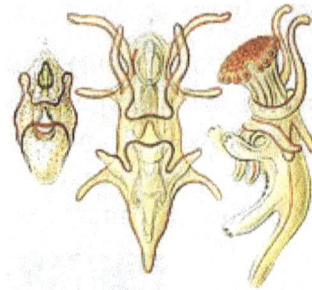

Starfish larvae are bilaterally symmetric, whereas the adults have fivefold symmetry

The echinoderms are important both biologically and geologically. Biologically, there are few other groupings so abundant in the biotic desert of the deep sea, as well as shallower oceans. Most echinoderms are able to regenerate tissue, organs, limbs, and reproduce asexually; in some cases, they can undergo complete regeneration from a single limb. Geologically, the value of echinoderms is in their ossified skeletons, which are major contributors to many limestone formations, and can provide valuable clues as to the geological environment. They were the most used species in regenerative research in the 19th and 20th centuries. Further, it is held by some scientists that the radiation of echinoderms was responsible for the Mesozoic Marine Revolution.

Echinoderm literally means "spiny skin", as this water melon sea urchin illustrates

Sea cucumbers filter feed on plankton and suspended solids

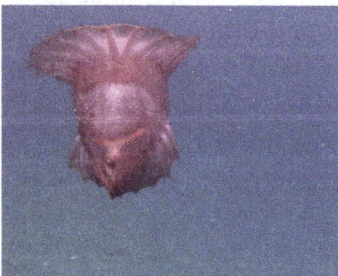

Benthopelagic sea cucumbers can lift off the seafloor and journey as much as 1,000 m (3,300 ft) up the water column

The ochre sea star was the first keystone predator to be studied. They limit mussels which can overwhelm intertidal communities.

Aside from the hard-to-classify *Arkarua* (a Precambrian animal with echinoderm-like pentamerous radial symmetry), the first definitive members of the phylum appeared near the start of the Cambrian.

Marine Molluscs

Molluscs (Latin for *soft*) form a phylum with about 85,000 extant recognized species. By species count they are the largest marine phylum, comprising about 23% of all the named marine organisms. Molluscs have more varied forms than other invertebrate phylums. They are highly diverse, not just in size and in anatomical structure, but also in behaviour and in habitat. The majority of species still live in the oceans, from the seashores to the abyssal zone, but some form a significant part of the freshwater fauna and the terrestrial ecosystems.

Reconstruction of an ammonite, a highly successful early cephalopod
that first appeared in the Devonian (about 400 mya).
They became extinct during the same extinction event that killed the land dinosaurs (about 66 mya).

The mollusc phylum is divided into 9 or 10 taxonomic classes, two of which are extinct. These classes include gastropods, bivalves and cephalopods, as well as other lesser-known but distinctive classes. Gastropods with protective shells are referred to as snails, whereas gastropods without protective shells are referred to as slugs. Gastropods are by far the most numerous molluscs in terms of classified species, accounting for 80% of the total. Bivalves include clams, oysters, cockles, mussels, scallops, and numerous other families. There are about 8,000 marine bivalves species (including brackish water and estuarine species), and about 1,200 freshwater species. Cephalopod include octopus, squid and cuttlefish. They are found in all oceans, and neurologically are the most advanced of the invertebrates. About 800 living species of marine cephalopods have been identified, and an estimated 11,000 extinct taxa have been described. There are no fully freshwater cephalopods.

The nautilus is a living fossil little changed since it evolved 500 million years ago as one of the first cephalopods.

Marine gastropods are sea snails or sea slugs. This nudibranch is a sea slug.

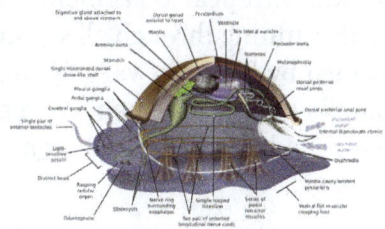

Generalized or *hypothetical ancestral mollusc*

Molluscs have such diverse shapes that many textbooks base their descriptions of molluscan anatomy on a generalized or *hypothetical ancestral mollusc*. This generalized mollusc is unsegmented and bilaterally symmetrical with an underside consisting of a single muscular foot. Beyond that it has three further key features. Firstly, it has a muscular cloak called a mantle covering its viscera and containing a significant cavity used for breathing and excretion. A shell secreted by the mantle covers the upper surface. Secondly (apart from bivalves) it has a rasping tongue called a radula used for feeding. Thirdly, it has a nervous system including a complex digestive system using microscopic, muscle-powered hairs called cilia to exude mucus. The generalized mollusc has two paired nerve cords (three in bivalves). The brain, in species that have one, encircles the esophagus. Most molluscs have eyes and all have sensors detecting chemicals, vibrations, and touch. The simplest type of molluscan reproductive system relies on external fertilization, but more complex variations occur. All produce eggs, from which may emerge trochophore larvae, more complex veliger larvae, or miniature adults. The depiction is rather similar to modern monoplacophorans, and some suggest it may resemble very early molluscs.

Good evidence exists for the appearance of marine gastropods, cephalopods and bivalves in the Cambrian period 541 to 485.4 million years ago. However, the evolutionary history both of molluscs' emergence from the ancestral Lophotrochozoa and of their diversification into the well-known living and fossil forms are still subjects of vigorous debate among scientists.

Marine Arthropods

Arthropods (Greek for *jointed feet*) have an exoskeleton (external skeleton), a segmented body, and jointed appendages (paired appendages). They form a phylum which includes insects, arachnids, myriapods, and crustaceans. Arthropods are characterized by their jointed limbs and cuticle made of chitin, often mineralised with calcium carbonate. The arthropod body plan consists of segments, each with a pair of appendages. The rigid cuticle inhibits growth, so arthropods replace it periodically by moulting. Their versatility has enabled them to become the most species-rich members of all ecological guilds in most environments.

Marine arthropods range in size from the microscopic crustacean *Stygotantulus* to the Japanese spider crab. Arthropods' primary internal cavity is a hemocoel, which accommodates their internal organs, and through which their haemolymph - analogue of blood - circulates; they have open circulatory systems. Like their exteriors, the internal organs of arthropods are generally built of repeated segments. Their nervous system is "ladder-like", with paired ventral nerve cords running through all segments and forming paired ganglia in each segment. Their heads are formed by fusion of varying numbers of segments, and their brains are formed by fusion of the ganglia of these segments and encircle the esophagus. The respiratory and excretory systems of arthropods vary, depending as much on their environment as on the subphylum to which they belong.

Their vision relies on various combinations of compound eyes and pigment-pit ocelli: in most species the ocelli can only detect the direction from which light is coming, and the compound eyes are the main source of information, but the main eyes of spiders are ocelli that can form images and, in a few cases, can swivel to track prey. Arthropods also have a wide range of chemical and mechanical sensors, mostly based on modifications of the many setae (bristles) that project through their cuticles. Arthropods' methods of reproduction and development are diverse; all terrestrial species use internal fertilization, but this is often by indirect transfer of the sperm via an appendage or the ground, rather than by direct injection. Marine species all lay eggs and use either internal or external fertilization. Arthropod hatchlings vary from miniature adults to grubs that lack jointed limbs and eventually undergo a total metamorphosis to produce the adult form.

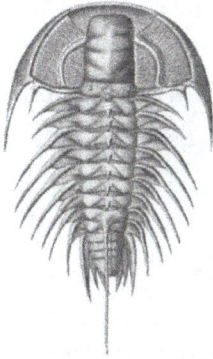

Trilobites, now extinct, roamed oceans for 270 million years.

Horseshoe crab, a living fossil arthropod from 450 million years ago

Many crustaceans are very small, like this tiny amphipod, and make up a significant part of the ocean's zooplankton

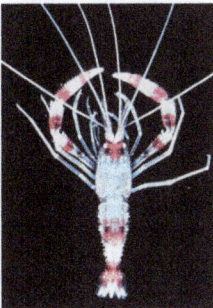

The banded cleaner shrimp is a crustacean common in the tropics.

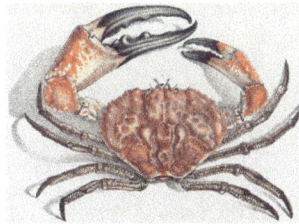

The Tasmanian giant crab is long-lived and slow-growing, making it vulnerable to overfishing.

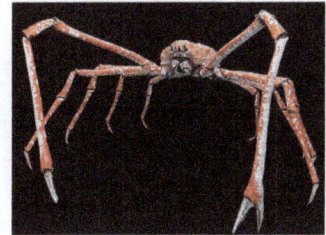

The Japanese spider crab has the longest leg span of any arthropod.

The evolutionary ancestry of arthropods dates back to the Cambrian period. The group is generally regarded as monophyletic, and many analyses support the placement of arthropods with cyclo-neuralians (or their constituent clades) in a superphylum Ecdysozoa. Overall however, the basal relationships of Metazoa are not yet well resolved. Likewise, the relationships between various arthropod groups are still actively debated.

Other Phyla

- Tardigrade, Lobopodia, (Onychophora)

- Non-craniate (non-vertebrate) chordates: Cephalochordate, Tunicata and *Haikouella*. These invertebrates are close relatives of the vertebrates.

Fluorescent-colored sea squirts, *Rhopalaea crassa*. Tunicates may provide clues to vertebrate (and therefore human) ancestry.

Salp chain

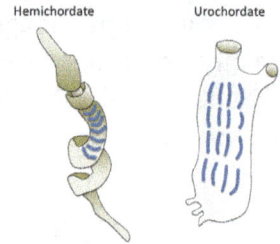

Gill slits in an acorn worm (left) and tunicate (right)

Minerals from Sea Water

There are a number of marine invertebrates that use minerals that are present in the sea in such minute quantities that they were undetectable until the advent of spectroscopy. Vanadium is concentrated by some tunicates for use in their blood cells to a level ten million times that of the surrounding seawater. Other tunicates similarly concentrate niobium and tantalum. Lobsters use copper in their respiratory pigment hemocyanin, despite the proportion of this metal in seawater being minute. Although these elements are present in vast quantities in the ocean, their extraction by man is not economic.

Echinoderm

Echinoderm is the common name given to any member of the phylum Echinodermata (from Ancient Greek, *echinos* – "hedgehog" and *derma* – "skin") of marine animals. The adults are recognizable by their (usually five-point) radial symmetry, and include such well-known animals as sea stars, sea urchins, sand dollars, and sea cucumbers, as well as the sea lilies or "stone lilies". Echinoderms are found at every ocean depth, from the intertidal zone to the abyssal zone. The phylum contains about 7000 living species, making it the second-largest grouping of deuterostomes (a superphylum), after the chordates (which include the vertebrates, such as birds, fishes, mammals, and reptiles). Echinoderms are also the largest phylum that has no freshwater or terrestrial (land-based) representatives.

Aside from the hard-to-classify *Arkarua* (a Precambrian animal with echinoderm-like pentamerous radial symmetry), the first definitive members of the phylum appeared near the start of the Cambrian. One group of Cambrian echinoderms, the cinctans (Homalozoa), which are close to the base of the echinoderm origin, have been found to possess external gills used for filter feeding, like chordata and hemichordata.

The echinoderms are important both ecologically and geologically. Ecologically, there are few other groupings so abundant in the biotic desert of the deep sea, as well as shallower oceans. Most echinoderms are able to regenerate tissue, organs, limbs, and reproduce asexually; in some cases, they can undergo complete regeneration from a single limb. Geologically, the value of echinoderms is in their ossified skeletons, which are major contributors to many limestone formations, and can provide valuable clues as to the geological environment. They were the most used species in regenerative research in the 19th and 20th centuries. Further, it is held by some scientists that the radiation of echinoderms was responsible for the Mesozoic Marine Revolution.

Taxonomy and Evolution

Along with the chordates and hemichordates, echinoderms are deuterostomes, one of the two major divisions of the bilaterians, the other being the protostomes. During the early development of the embryo, in deuterostomes, the blastopore (the first opening to form) becomes the anus whereas in the protostomes, it becomes the mouth. In deuterostomes, the mouth develops at a later stage, at the opposite end of the blastula from the blastopore, and a gut forms connecting the two. The larvae of echinoderms have bilateral symmetry but this is lost during metamorphosis when their bodies are reorganised and develop the characteristic radial symmetry of the echinoderm, typically pentamerism. The characteristics of adult echinoderms are the possession of a water vascular system with external tube feet and a calcareous endoskeleton consisting of ossicles connected by a mesh of collagen fibres.

The Ordovician cystoid *Echinosphaerites* from northeastern Estonia

Fossil crinoid crowns

There are a total of about 7,000 extant species of echinoderm as well as about 13,000 extinct species. They are found in habitats ranging from shallow intertidal areas to abyssal depths. Two main subdivisions are traditionally recognised: the more familiar motile Eleutherozoa, which encompasses the Asteroidea (starfish, 1,745 recent species), Ophiuroidea (brittle stars, 2,300 species), Echinoidea (sea urchins and sand dollars, 900 species) and Holothuroidea (sea cucumbers, 1,430 species); and the Pelmatozoa, some of which are sessile while others move around. These consist of the Crinoidea (feather stars and sea lilies, 580 species) and the extinct blastoids and Paracrinoids. A fifth class of Eleutherozoa consisting of just three species, the Concentricycloidea (sea daisies), were recently merged into the Asteroidea. The fossil record includes a large number of other classes which do not appear to fall into any extant crown group.

All echinoderms are marine and nearly all are benthic. The oldest known echinoderm fossil may be *Arkarua* from the Precambrian of Australia. It is a disc-like fossil with radial ridges on the rim and a five-pointed central depression marked with radial lines. However, no stereom or internal structure showing a water vascular system is present and the identification is inconclusive.

The first universally accepted echinoderms appear in the Lower Cambrian period, asterozoans appeared in the Ordovician and the crinoids were a dominant group in the Paleozoic. Echinoderms left behind an extensive fossil record. It is hypothesised that the ancestor of all echinoderms was a simple, motile, bilaterally symmetrical animal with a mouth, gut and anus. This ancestral stock adopted an attached mode of life and suspension feeding, and developed radial symmetry as this was more advantageous for such an existence. The larvae of all echinoderms are even now bilaterally symmetrical and all develop radial symmetry at metamorphosis. The starfish and crinoids still attach themselves to the seabed while changing to their adult form.

The first echinoderms later gave rise to free-moving groups. The evolution of endoskeletal plates with stereom structure and of external ciliary grooves for feeding were early echinoderm developments. The Paleozoic echinoderms were globular, attached to the substrate and were orientated with their oral surfaces upwards. The fossil echinoderms had ambulacral grooves extending down the side of the body, fringed on either side by brachioles, structures very similar to the pinnules of a modern crinoid. It seems probable that the mouth-upward orientation is the primitive state and that at some stage, all the classes of echinoderms except the crinoids reversed this to become mouth-downward. Before this happened, the podia probably had a feeding function as they do in the crinoids today. Their locomotor function came later, after the re-orientation of the mouth when the podia were in contact with the substrate for the first time.

Anatomy and Physiology

Echinoderms evolved from animals with bilateral symmetry. Although adult echinoderms possess pentaradial, or five-sided, symmetry, echinoderm larvae are ciliated, free-swimming organisms that organize in bilateral symmetry which makes them look like embryonic chordates. Later, the left side of the body grows at the expense of the right side, which is eventually absorbed. The left side then grows in a pentaradially symmetric fashion, in which the body is arranged in five parts around a central axis.

Echinoderms exhibit secondary radial symmetry in portions of their body at some stage of life. This, however, is an adaptation to their sessile existence. They developed from other members of the Bilateria and exhibit bilateral symmetry in their larval stage. Many crinoids and some seastars exhibit symmetry in multiples of the basic five, with starfish such as *Labidiaster annulatus* known to possess up to fifty arms, and the sea-lily *Comaster schlegelii* having two hundred.

Skin and Skeleton

A brittle star, *Ophionereis reticulata*

A sea cucumber from Malaysia

Starfish exhibit a wide range of colours

Strongylocentrotus purpuratus, a well-armoured sea urchin

Echinoderms have a mesodermal skeleton composed of calcareous plates or ossicles. Each one of these, even the articulating spine of a sea urchin, is composed mineralogically of a crystal of calcite. If solid, these would form a heavy skeleton, so they have a sponge-like porous structure known as stereom. Ossicles may be fused together, as in the test of sea urchins, or may articulate with each other as in the arms of sea stars, brittle stars and crinoids. The ossicles may be flat plates or bear external projections in the form of spines, granules or warts and they are supported by a tough epidermis (skin). Skeletal elements are also deployed in some specialized ways, such as the "Aristotle's lantern" mouthparts of sea urchins used for grinding, the supportive stalks of crinoids and the structural "lime ring" of sea cucumbers.

Despite the robustness of the individual skeletal modules complete skeletons of starfish, brittle stars and crinoids are rare in the fossil record. This is because they quickly disarticulate (disconnect from each other) once the encompassing skin rots away, and in the absence of tissue there is nothing to hold the plates together. The modular construction is a result of the growth system employed by echinoderms, which adds new segments at the centre of the radial limbs, pushing the existing plates outwards and lengthening the arms. Sea urchins on the other hand are often well preserved in chalk beds or limestone. During fossilization, the cavities in the stereom are filled in with calcite that is in crystalline continuity with the surrounding material. On fracturing such rock, distinctive cleavage patterns can be seen and sometimes even the intricate internal and external structure of the test.

The epidermis consists of cells responsible for the support and maintenance of the skeleton, as well as pigment cells, mechanoreceptor cells (which detect motion on the animal's surface), and sometimes gland cells which secrete sticky fluids or even toxins. The varied and often vivid colours of echinoderms are produced by the action of skin pigment cells. These are produced by a variable combination of coloured pigments, such as the dark melanin, red carotinoids, and carotene proteins, which can be blue, green, or violet. These may be light-sensitive, and as a result many echinoderms change appearance completely as night falls. The reaction can happen quickly — the sea urchin *Centrostephanus longispinus* changes from jet black to grey-brown in just fifty minutes when exposed to light.

One characteristic of most echinoderms is a special kind of tissue known as "catch connective tissue". This collagenous material can change its mechanical properties in a few seconds or minutes through nervous control rather than by muscular means. This tissue enables a starfish to change from moving flexibly around the seabed to becoming rigid while prying open a bivalve mollusc or preventing itself from being extracted from a crevice. Similarly, sea urchins can lock their normally mobile spines rigidly as a defensive mechanism when attacked.

The Water Vascular System

Echinoderms possess a unique water vascular system. This is a network of fluid-filled canals derived from the coelom (body cavity) that function in gas exchange, feeding, sensory reception and locomotion. This system varies between different classes of echinoderm but typically opens to the exterior through a sieve-like madreporite on the aboral (upper) surface of the animal. The madreporite is linked to a slender duct, the stone canal, which extends to a ring canal that encircles the mouth or oesophagus. From this, radial canals extend along the arms of asteroids and adjoin the test in the ambulacral areas of echinoids. Short lateral canals branch off the radial canals, each one ending in an ampulla. Part of the ampulla can protrude through a pore (or a pair of pores in sea urchins) to the exterior and is known as a podium or tube feet. The water vascular system assists with the distribution of nutrients throughout the animal's body and is most obviously expressed in the tube feet which can be extended or contracted by the redistribution of fluid between the foot and the internal sac.

The organization of the system is somewhat different in ophiuroids where the medreporite may be on the oral surface and the podia lack suckers. In holothuroids, the podia may be reduced or absent and the madreporite opens into the body cavity so that the circulating liquid is coelomic fluid rather than sea water. The arrangements in crinoids is similar to asteroids but the tube feet lack suckers and are used to pass food particles captured by the arms towards the central mouth. In the asteroids, the same wafting motion is employed to move the animal across the ground. Sea urchins use their feet to prevent the larvae of encrusting organisms from settling on their surfaces; potential settlers are moved to the urchin's mouth and eaten. Some burrowing sea stars extend their elongated dorsal tube feet to the surface of the sand or mud above and use them to absorb oxygen from the water column.

Other Organs

Echinoderms possess a simple digestive system which varies according to the animal's diet. Starfish are mostly carnivorous and have a mouth, oesophagus, two-part stomach, intestine and rectum, with the anus located in the centre of the aboral body surface. In many species, the large cardiac stomach can be everted and digest food outside the body. In other species, whole food items such as molluscs may be ingested. Brittle stars have a blind gut with no intestine or anus. They have varying diets and expel food waste through their mouth. Sea urchins are herbivores and use their specialised mouthparts to graze, tear and chew algae and sometimes other animal or vegetable material. They have an oesophagus, a large stomach and a rectum with the anus at the apex of the test. Sea cucumbers are mostly detritivores, sorting through the sediment with their buccal tentacles which are modified tube feet. Sand and mud accompanies their food through their simple gut which has a long coiled intestine and a capacious cloaca. Crinoids are passive suspension feeders, catching plankton with their outstretched arms. Boluses of mucus-trapped food are passed to the mouth which is linked to the anus by a loop consisting of a short oesophagus and longer intestine.

The coelomic cavities of echinoderms are complex. Aside from the water vascular system, echinoderms have a haemal coelom (or haemal system, the "haemal" being a misnomer), a perivisceral coelom, a gonadal coelom and often also a perihaemal coelom (or perihaemal system). During development, echinoderm coelom is divided in metacoel, mesocoel and protocoel (also called somatocoel, hydrocoel and axocoel, respectively). The water vascular system, haemal system and perihaemal system form the tubular coelomic system. Echinoderms are an exception having both a coelomic circulatory system (i.e., the water vascular system) and a haemal circulatory system (i.e., the haemal and perihaemal systems).

Haemal and perihaemal systems are derived from the coelom and form an open and reduced circulatory system. This usually consists of a central ring and five radial vessels. There is no true heart and the blood often lacks any respiratory pigment. Gaseous exchange occurs via dermal branchae or papulae in starfish, genital bursae in brittle stars, peristominal gills in sea urchins and cloacal trees in sea cucumbers. Exchange of gases also takes place through the tube feet. Echinoderms lack specialized excretory (waste disposal) organs and so nitrogenous waste, chiefly in the form of ammonia, diffuses out through the respiratory surfaces.

The coelomic fluid contains the coelomocytes, or immune cells. There are several types of immune cells, which vary among classes and species. All classes possess a type of phagocytic amebocyte, which engulf invading particles and infected cells, aggregate or clot, and may be involved in cytotoxicity. These cells are usually larger and granular, and are suggested to be a main line of defense against potential pathogens. Depending on the class, echinoderms may have spherule cells (for cytotoxicity, inflammation, and anti-bacterial activity), vibratile cells (for coelomic fluid movement and clotting), and crystal cells (potential osmoregulatory cells in sea cucumbers),. The coelomocytes also secrete Anti-Microbial Peptides (AMPs) against bacteria, and have a set of lectins and complement proteins as part of an innate immune system that is still being characterized.

Echinoderms have a simple radial nervous system that consists of a modified nerve net consisting of interconnecting neurons with no central brain, although some do possess ganglia. Nerves radiate from central rings around the mouth into each arm or along the body wall; the branches of these nerves coordinate the movements of the organism and the synchronisation of the tube feet. Starfish have sensory cells in the epithelium and have simple eyespots and touch-sensitive tentacle-like tube feet at the tips of their arms. Sea urchins have no particular sense organs but do have statocysts that assist in gravitational orientation, and they have sensory cells in their epidermis, particularly in the tube feet, spines and pedicellariae. Brittle stars, crinoids and sea cucumbers in general do not have sensory organs but some burrowing sea cucumbers of the order Apodida have a single statocyst adjoining each radial nerve and some have an eyespot at the base of each tentacle.

The gonads occupy much of the body cavities of sea urchins and sea cucumbers, while the less voluminous crinoids, brittle stars and starfish have two gonads in each arm. While the ancestral condition is considered to be the possession of one genital aperture, many organisms have multiple gonopores through which eggs or sperm may be released.

Regeneration

Many echinoderms have remarkable powers of regeneration. Many species routinely autotomize and regenerate arms and viscera. Sea cucumbers often discharge parts of their internal organs if they perceive themselves to be threatened. The discharged organs and tissues are regenerated over the course of several months. Sea urchins are constantly replacing spines lost through damage.

Sea stars and sea lilies readily lose and regenerate their arms. In most cases, a single severed arm cannot grow into a new starfish in the absence of at least part of the disc. However, in a few species a single arm can survive and develop into a complete individual and in some species, the arms are intentionally detached for the purpose of asexual reproduction. During periods when they have lost their digestive tracts, sea cucumbers live off stored nutrients and absorb dissolved organic matter directly from the water.

Sunflower star regenerating several arms

The regeneration of lost parts involves both epimorphosis and morphallaxis. In epimorphosis stem cells—either from a reserve pool or those produced by dedifferentiation—form a blastema and generate new tissues. Morphallactic regeneration involves the movement and remodelling of existing tissues to replace lost parts. Direct transdifferentiation of one type of tissue to another during tissue replacement is also observed.

The robust larval growth is responsible for many echinoderms being used as popular model organisms in developmental biology.

Reproduction

Sexual Reproduction

Echinoderms become sexually mature after approximately two to three years, depending on the species and the environmental conditions. They are nearly all gonochoric, though a few species are hermaphroditic. The eggs and sperm cells are typically released into open water, where fertilization takes place. The release of sperm and eggs is synchronised in some species, usually with regard to the lunar cycle. In other species, individuals may aggregate during the reproductive season, thereby increasing the likelihood of successful fertilisation. Internal fertilisation has currently been observed in three species of sea star, three brittle stars and a deep water sea cucumber. Even at abyssal depths, where no light penetrates, synchronisation of reproductive activity in echinoderms is surprisingly frequent.

Some echinoderms brood their eggs. This is especially common in cold water species where planktonic larvae might not be able to find sufficient food. These retained eggs are usually few in number and are supplied with large yolks to nourish the developing embryos. In starfish, the female may carry the eggs in special pouches, under her arms, under her arched body or even in her cardiac stomach. Many brittle stars are hermaphrodites. Egg brooding is quite common and usually takes place in special chambers on their oral surfaces, but sometimes the ovary or coelom is used. In these starfish and brittle stars, direct development without passing through a bilateral larval stage usually takes place. A few sea urchins and one species of sand dollar carry their eggs in cavities, or near their anus, holding them in place with their spines. Some sea cucumbers use their buccal tentacles to transfer their eggs to their underside or back where they are retained. In a very small number of species, the eggs are retained in the coelom where they develop viviparously, later emerging through ruptures in the body wall. In some species of crinoid, the embryos develop in special breeding bags, where the eggs are held until sperm released by a male happens to find them.

Asexual Reproduction

One species of seastar, *Ophidiaster granifer*, reproduces asexually by parthenogenesis. In certain other asterozoans, the adults reproduce asexually for a while before they mature after which time they reproduce sexually. In most of these species, asexual reproduction is by transverse fission with the disc splitting in two. Regrowth of both the lost disc area and the missing arms occur so that an individual may have arms of varying lengths. Though in most species at least part of the disc is needed for complete regeneration, in a few species of sea stars, a single severed arm can grow into a complete individual over a period of several months. In at least some of these species, they actively use this as a method of asexual reproduction. A fracture develops on the lower surface of the arm and the arm pulls itself free from the body which holds onto the substrate during the process. During the period of regrowth, they have a few tiny arms and one large arm, thus often being referred to as "comets".

Asexual reproduction by transverse fission has also been observed in adult sea cucumbers. *Holothuria parvula* uses this method frequently, an individual splitting into two a little in front of the midpoint. The two halves each regenerate their missing organs over a period of several months but the missing genital organs are often very slow to develop.

The larvae of some echinoderm species are capable of asexual reproduction. This has long been known to occur among starfish and brittle stars but has been more recently observed in a sea cucumber, a sand dollar and a sea urchin. These species belong to four of the major classes of echinoderms except crinozoans (as of 2011). Asexual reproduction in the planktonic larvae occurs through numerous modes. They may autotomise parts that develop into secondary larvae, grow buds or undergo paratomy. The parts that are autotomised or the buds may develop directly into fully formed larvae or may develop through a gastrula or even a blastula stage. The parts that develop into the new larvae vary from the preoral hood (a mound like structure above the mouth), the side body wall, the postero-lateral arms or their rear ends.

The process of cloning is a cost borne by the larva both in resources as well as in development time. Larvae have been observed to undergo this process when food is plentiful or temperature conditions are optimal. It has also been suggested that cloning may occur to make use of the tissues that are normally lost during metamorphosis. Recent research has shown that the larvae of some sand dollars clone themselves when they detect predators (by sensing dissolved fish mucus). Asexual reproduction produces many smaller larvae that escape better from planktivorous fish.

Larval Development

The development of an echinoderm begins with a bilaterally symmetrical embryo, with a coeloblastula developing first. Gastrulation marks the opening of the "second mouth" that places echinoderms within the deuterostomes, and the mesoderm, which will host the skeleton, migrates inwards. The secondary body cavity, the coelom, forms by the partitioning of three body cavities. The larvae are mostly planktonic but in some species the eggs are retained inside the female and in some, the larvae are also brooded by the female.

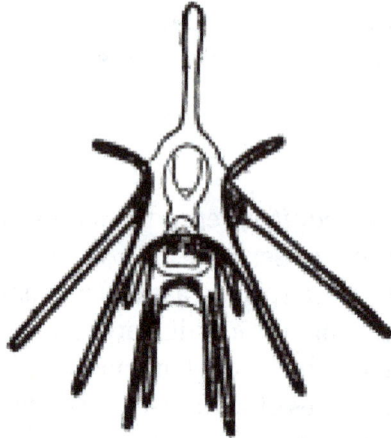

An echinopluteus larva with larval arms

The larvae of echinoderms pass through a number of stages and these have specific names derived from the taxonomic names of the adults or from their appearance. For example, a sea urchin has an 'echinopluteus' larva while a brittle star has an 'ophiopluteus' larva. A starfish has a 'bipinnaria' larva but this later develops into a multi-armed 'brachiolaria' larva. A sea cucumber larva is an 'auricularia' while a crinoid one is a 'vitellaria'. All these larvae are bilaterally symmetrical and have bands of cilia with which they swim and some, usually known as 'pluteus' larvae, have arms. When fully developed they settle on the seabed to undergo metamorphosis and the larval arms and gut degenerate. The left hand side of the larva develops into the oral surface of the juvenile while the right side becomes the aboral surface. At this stage the bilateral symmetry is lost and radial symmetry develops.

The planktotrophic larva is considered to be the ancestral larval type for echinoderms but after 500 million years of larval evolution, about 68% of species whose development is known have a lecithotrophic larval type. The provision of a yolk-sac means that smaller numbers of eggs are produced, the larvae have a shorter development period, smaller dispersal potential but a greater chance of survival. There seems to be an evolutionary trend towards a "lower-risk–lower-gain" strategy of direct development.

Distribution and Habitat

Echinoderms are globally distributed in almost all depths, latitudes and environments in the ocean. They reach highest diversity in reef environments but are also widespread on shallow shores, around the poles — refugia where crinoids are at their most abundant — and throughout the deep ocean, where bottom-dwelling and burrowing sea cucumbers are common — sometimes accounting for up to 90% of organisms. While almost all echinoderms are benthic — that is, they live on the sea floor — some sea-lilies can swim at great velocity for brief periods of time, and a few deep-sea

sea cucumbers are fully floating. Some crinoids are pseudo-planktonic, attaching themselves to floating logs and debris, although this behaviour was exercised most extensively in the Paleozoic, before competition from such organisms as barnacles restricted the extent of the behaviour.

The larvae of echinoderms, especially starfish and sea urchins, are pelagic, and with the aid of ocean currents can be transported for great distances, reinforcing the global distribution of the phylum.

Mode of Life

Locomotion

Echinoderms primarily use their tube feet to move about but some sea urchins also use their spines. The tube feet typically have a tip shaped like a suction pad in which a vacuum can be created by contraction of muscles. This along with some stickiness provided by the secretion of mucus provides adhesion. Waves of tube feet contractions and relaxations move along the adherent surface and the animal moves slowly along.

Brittle stars are the most agile of the echinoderms, raising their discs and taking strides when moving. The two forward arms grip the substrate with their tube feet, the two side arms "row", the hindermost arm trails and the animal moves in jerks. The arm spines provide traction and when moving among objects, the supple arms can coil around things. A few species creep around on pointed tube feet. Starfish extend their tube feet in the intended direction of travel and grip the substrate by suction, after which the feet are drawn backwards. The movement of multiple tube feet, coordinated in waves, moves the animal forward, but progress is slow. Some burrowing starfish have points rather than suckers on their tube feet and they are able to "glide" across the seabed at a faster rate.

Sea urchins use their tube feet to move around in a similar way to starfish. Some also use their articulated spines to push or lever themselves along or lift their oral surfaces off the substrate. If a sea urchin is overturned, it can extend its tube feet in one ambulacral area far enough to bring them within reach of the substrate and then successively attach feet from the adjoining area until it is righted. Some species bore into rock and they usually do this by grinding away at the surface with their mouthparts.

Sea cucumbers are generally sluggish animals. Many can move on the surface or burrow through sand or mud using peristaltic movements and some have short tube feet on their under surface with which they can creep along in the manner of a starfish. Some species drag themselves along by means of their buccal tentacles while others can expand and contract their body or rhythmically flex it and "swim". Many live in cracks, hollows and burrows and hardly move at all. Some deep water species are pelagic and can float in the water with webbed papillae forming sails or fins.

The majority of crinoids are motile but the sea lilies are sessile and attached to hard substrates by stalks. These stems can bend and the arms can roll and unroll and that is about the limit of the sea lily's movement, although a few species can relocate themselves on the seabed by crawling. The sea feathers are unattached and usually live in crevices, under corals or inside sponges with their arms the only visible part. Some sea feathers emerge at night and perch themselves on nearby eminences to better exploit the food-bearing current. Many species can "walk" across the seabed, raising their body with the help of their arms. Many can also swim with their arms but most are largely sedentary, seldom moving far from their chosen place of concealment.

Feeding

The modes of feeding vary greatly between the different echinoderm taxa. Crinoids and some brittle stars tend to be passive filter-feeders, enmeshing suspended particles from passing water; most sea urchins are grazers, sea cucumbers deposit feeders and the majority of starfish are active hunters.

Crinoids are suspension feeders and spread their arms wide to catch particles floating past. These are caught by the tube feet on the pinnules, moved into the ambulacral grooves, wrapped in mucus and conveyed to the mouth by the cilia lining the grooves. The exact dietary requirements of crinoids have been little researched but in the laboratory they can be fed with diatoms.

Basket stars are suspension feeders, raising their branched arms to collect zooplankton, while brittle stars use several methods of feeding, though usually one predominates. Some are suspension feeders, securing food particles with mucus strands, spines or tube feet on their raised arms. Others are scavengers and feeders on detritus. Others again are voracious carnivores and able to lasso their waterborne prey with a sudden encirclement by their flexible arms. The limbs then bend under the disc to transfer the food to the jaws and mouth.

Many sea urchins feed on algae, often scraping off the thin layer of algae covering the surfaces of rocks with their specialised mouthparts known as Aristotle's lantern. Other species devour smaller organisms, which they may catch with their tube feet. They may also feed on dead fish and other animal matter. Sand dollars may perform suspension feeding and feed on phytoplankton, detritus, algal pieces and the bacterial layer surrounding grains of sand.

Many sea cucumbers are mobile deposit or suspension feeders, using their buccal podia to actively capture food and then stuffing the particles individually into their buccal cavities. Others ingest large quantities of sediment, absorb the organic matter and pass the indigestible mineral particles through their guts. In this way they disturb and process large volumes of substrate, often leaving characteristic ridges of sediment on the seabed. Some sea cucumbers live infaunally in burrows, anterior-end down and anus on the surface, swallowing sediment and passing it through their gut. Other burrowers live anterior-end up and wait for detritus to fall into the entrances of the burrows or rake in debris from the surface nearby with their buccal podia.

Nearly all starfish are detritivores or carnivores, though a few are suspension feeders. Small fish landing on the upper surface may be captured by pedicilaria and dead animal matter may be scavenged but the main prey items are living invertebrates, mostly bivalve molluscs. To feed on one of these, the starfish moves over it, attaches its tube feet and exerts pressure on the valves by arching its back. When a small gap between the valves is formed, the starfish inserts part of its stomach into the prey, excretes digestive enzymes and slowly liquefies the soft body parts. As the adductor muscle of the shellfish relaxes, more stomach is inserted and when digestion is complete, the stomach is returned to its usual position in the starfish with its now liquefied bivalve meal inside it. The same everted stomach process is used by other starfish to feed on sponges, sea anemones, corals, detritus and algal films.

Defense Mechanisms

Despite their low nutrition value and the abundance of indigestible calcite, echinoderms are the prey of many organisms, such as crabs, sharks, sea birds and other echinoderms. Defensive strategies employed include the presence of spines, toxins, which can be inherent or delivered through the tube feet, and the discharge of sticky entangling threads by sea cucumbers. Although most echinoderm spines are blunt, those of the crown-of-thorns starfish are long and sharp and can cause a painful puncture wound as the epithelium covering them contains a toxin. Because of their catch connective tissue, which can change rapidly from a flaccid to a rigid state, echinoderms are very difficult to dislodge from crevices. Certain sea cucumbers have a cluster of cuvierian tubules which can be ejected as long sticky threads from their anus and entangle and permanently disable an attacker. Another defensive strategy sometimes adopted by sea cucumbers is to rupture the body wall and discharge the gut and internal organs. The animal has a great regenerative capacity and will regrow the lost parts later. Starfish and brittle stars may undergo autotomy when attacked, an arm becoming detached which may distract the predator for long enough for the animal to escape. Some starfish species can "swim" away from what may be danger, foregoing the regrowth by not losing limbs. It is not unusual to find starfish with arms of different sizes in various stages of regrowth.

Ecology

Echinoderms are numerous and relatively large invertebrates and play an important role in marine, benthic ecosystems. The grazing of sea urchins reduces the rate of colonization of bare rock by settling organisms but also keeps algae in check, thereby enhancing the biodiversity of coral reefs. The burrowing of sand dollars, sea cucumbers and some starfish stirs up the sediment and depletes the sea floor of nutrients. Their digging activities increases the depth to which oxygen can seep and allows a more complex ecological tier-system to develop. Starfish and brittle stars prevent the growth of algal mats on coral reefs, which might otherwise obstruct the filter-feeding constituent organisms. Some sea urchins can bore into solid rock and this bioerosion can destabilise rock faces and release nutrients into the ocean. Coral reefs are also bored into in this way but the rate of accretion of carbonate material is often greater than the erosion produced by the sea urchin. It has been estimated that echinoderms capture and sequester about 0.1 gigatonnes of carbon per year as calcium carbonate, making them important contributors in the global carbon cycle.

Echinoderms sometimes have large population swings which can cause marked consequences for ecosystems. An example is the change from a coral-dominated reef system to an alga-dominated one that resulted from the mass mortality of the tropical sea urchin *Diadema antillarum* in the Caribbean in 1983. Sea urchins are among the main herbivores on reefs and there is usually a fine balance between the urchins and the kelp and other algae on which they graze. A diminution of the numbers of predators (otters, lobsters and fish) can result in an increase in urchin numbers causing overgrazing of kelp forests with the result that an alga-denuded "urchin barren" forms. On the Great Barrier Reef, an unexplained increase in the numbers of crown-of-thorns starfish (*Acanthaster planci*), which graze on living coral tissue, has had considerable impact on coral mortality and coral reef biodiversity.

Echinoderms form part of the diet of many organisms such as bony fish, sharks, eider ducks, gulls, crabs, gastropod molluscs, sea otters, Arctic foxes and humans. Larger starfish prey on smaller ones and the great quantity of eggs and larvae produced form part of the zooplankton, consumed by many marine creatures. Crinoids are relatively free from predation. The body cavities of many sea cucumbers and some starfish provide a habitat for parasitic or symbiotic organisms including fish, crabs, worms and snails.

Use by Humans

In 2010, 373,000 tonnes of echinoderms were harvested, mainly for consumption. These were mainly sea cucumbers (158,000 tonnes) and sea urchins (73,000 tonnes).

Sea cucumbers are considered a delicacy in some countries of south east Asia; as such, they are in imminent danger of being over-harvested.

Popular species include the pineapple roller *Thelenota ananas* (*susuhan*) and the red *Holothuria edulis*. These and other species are colloquially known as *bêche de mer* or *trepang* in China and Indonesia. The sea cucumbers are boiled for twenty minutes and then dried both naturally and later over a fire which gives them a smoky tang. In China they are used as a basis for gelatinous soups and stews. Both male and female gonads of sea urchins are also consumed particularly in Japan, Peru, Spain and France. The taste is described as soft and melting, like a mixture of seafood and fruit. The quality is assessed by the colour which can range from light yellow to bright orange. At the present time, some trials of breeding sea uchins in order to try to compensate the overexploitation of this resource have been made.

The calcareous tests or shells of echinoderms are used as a source of lime by farmers in areas where limestone is unavailable and some are used in the manufacture of fish meal. Four thousand tons of the animals are used annually for these purposes. This trade is often carried out in conjunction with shellfish farmers, for whom the starfish pose a major threat by eating their cultured stock. Other uses for the starfish they recover include the manufacture of animal feed, composting and drying for the arts and craft trade.

Sea urchins are used in research, particularly as model organisms in developmental biology and ecotoxicology. *Strongylocentrotus purpuratus* and *Arbacia punctulata* are used for this purpose in embryological studies. The large size and the transparency of the eggs enables the observation of sperm cells in the process of fertilising ova. The arm regeneration potential of brittle stars is being studied in connection with understanding and treating neurodegenerative diseases in humans.

Sea Cucumber

Sea cucumbers are echinoderms from the class Holothuroidea. They are marine animals with a leathery skin and an elongated body containing a single, branched gonad. Sea cucumbers are found on the sea floor worldwide. The number of holothurian species worldwide is about 1,717 with the greatest number being in the Asia Pacific region. Many of these are gathered for human consumption and some species are cultivated in aquaculture systems. The harvested product is variously referred to as *trepang*, *bêche-de-mer* or *balate*. Sea cucumbers serve a useful role in the marine ecosystem as they help recycle nutrients, breaking down detritus and other organic matter after which bacteria can continue the degradation process. Like all echinoderms, sea cucumbers have an endoskeleton just below the skin, calcified structures that are usually reduced to isolated microscopic ossicles (or sclerietes) joined by connective tissue. In some species these can sometimes be enlarged to flattened plates, forming an armour. In pelagic species such as *Pelagothuria natatrix* (Order Elasipodida, family Pelagothuriidae), the skeleton is absent and there is no calcareous ring.

Thelenota ananas, a giant sea cucumber from the Indo-Pacific tropics.

The sea cucumbers are named after their resemblance to the fruit of the cucumber plant.

Overview

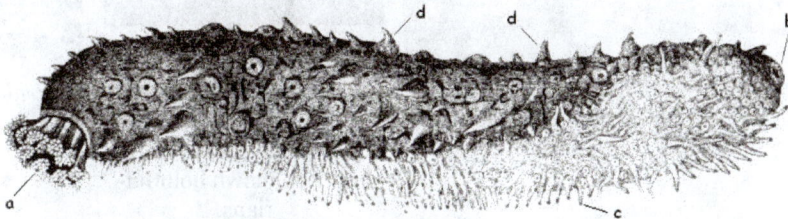

Sea cucumber : a -Tentacles, b - Cloaca, c - Ambulacral feet on the ventral side, d -Papillae on the back.

Most sea cucumbers, as their name suggests, have a soft and cylindrical body, more or less lengthened, rounded off and occasionally fat in the extremities, and generally without solid appendages. Their shape ranges from almost spherical for "sea apples" (genus *Pseudocolochirus*) to serpent-like for Apodida or the classic sausage-shape, while others resemble caterpillars. "The mouth is surrounded by tentacles, which can be pulled back inside the animal." Holothurians measure generally between 10 and 30 centimeters long, with extremes of some millimeters for *Rhabdomolgus ruber* and up to more than 3 meters for *Synapta maculata*. The largest American species, *Holothuria floridana*, which abounds just below low-water mark on the Florida reefs, has a volume of well over 500 cubic centimeters (31 cu in), and 25–30 cm (10–12 in) long. Most possess five rows of tube feet (called "podia"), but Apodida lacks these and moves by crawling; the podia can be of smooth aspect or provided with fleshy appendages (like *Thelenota ananas*). The podia on the dorsal surface generally have no locomotive role, and are transformed into papillae. At one of the extremities opens a rounded mouth, generally surrounded with a crown of tentacles which can be very complex in some species (they are in fact modified podia); the anus is postero-dorsal.

Holothurians do not look like other echinoderms at first glance, because of their tubular body, without visible skeleton nor hard appendixes. Furthermore, the fivefold symmetry, classical for echinoderms, although preserved structurally, is doubled here by a bilateral symmetry which makes them look like chordates. However, a central symmetry is still visible in some species through five 'radii', which extend from the mouth to the anus (just like for sea urchins), on which the tube feet are attached. There is thus no "oral" or "aboral" face as for sea stars and other echinoderms, but the animal stands on one of its sides, and this face is called *trivium* (with three rows of tube feet), while the dorsal face is named *bivium*. A remarkable feature of these animals is the "catch" collagen that forms their body wall. This can be loosened and tightened at will, and if the animal wants to squeeze through a small gap, it can essentially liquefy its body and pour into the space. To keep itself safe in these crevices and cracks, the sea cucumber will hook up all its collagen fibers to make its body firm again.

The most common way to separate the subclasses is by looking at their oral tentacles. Order Apodida have a slender and elongate body lacking tube feet, with up to 25 simple or pinnate oral tentacles. Aspidochirotida are the most common sea cucumbers encountered, with a strong body and 10–30 leaf like or shield like oral tentacles. Dendrochirotida are filter-feeders, with plump bodies and 8–30 branched oral tentacles (which can be extremely long and complex).

Details of the mouth with its tentacles.

Synaptula lamperti lives on sponges (here in Indonesia).

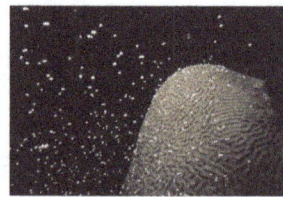

The king sea cucumber (*Thelenota anax*, Stichopodidae family) is one of the heaviest known holothurians.

Chiridota heheva, abyssal species.

Holothuria leucospilota

Isostichopus badionotus

Thelenota rubralineata

Holothuria fuscopunctata

Anatomy

Sea cucumbers are typically 10 to 30 cm (3.9 to 11.8 in) in length, although the smallest known species are just 3 mm (0.12 in) long, and the largest can reach 3 meters (9.8 ft). The body ranges from almost spherical to worm-like, and lacks the arms found in many other echinoderms, such as starfish. The anterior end of the animal, containing the mouth, corresponds to the oral pole of other echinoderms (which, in most cases, is the underside), while the posterior end, containing the anus, corresponds to the aboral pole. Thus, compared with other echinoderms, sea cucumbers can be said to be lying on their side.

Conspicuous Sea Cucumber, Coconut Island, Hawaii

Body Plan

The body of a holothurian is roughly cylindrical. It is radially symmetrical along its longitudinal axis, and has weak bilateral symmetry transversely with a dorsal and a ventral surface. As in other Echinozoans, there are five ambulacra separated by five ambulacral grooves, the interambulacra. The ambulacral grooves bear four rows of tube feet but these are diminished in size or absent in some holothurians, especially on the dorsal surface. The two dorsal ambulacra make up the bivium while the three ventral ones are known as the trivium.

At the anterior end, the mouth is surrounded by a ring of tentacles which are usually retractable into the mouth. These are modified tube feet and may be simple, branched or arborescent. They are known as the introvert and posterior to them there is an internal ring of large calcareous ossicles. Attached to this are five bands of muscle running internally longitudinally along the ambulacra. There are also circular muscles, contraction of which cause the animal to elongate and the introvert to extend. Anterior to the ossicles lie further muscles, contraction of which cause the introvert to retract.

The body wall consists of an epidermis and a dermis and contains smaller calcareous ossicles, the types of which are characteristics which help to identify different species. Inside the body wall is the coelom which is divided by three longitudinal mesenteries which surround and support the internal organs.

Digestive System

A sea cucumber atop gravel, feeding

A pharynx lies behind the mouth and is surrounded by a ring of ten calcareous plates. In most sea cucumbers, this is the only substantial part of the skeleton, and it forms the point of attachment for muscles that can retract the tentacles into the body for safety as for the main muscles of the body wall. Many species possess an oesophagus and stomach, but in some the pharynx opens directly into the intestine. The intestine is typically long and coiled, and loops through the body three times before terminating in a cloacal chamber, or directly as the anus.

Nervous System

Sea cucumbers have no true brain. A ring of neural tissue surrounds the oral cavity, and sends nerves to the tentacles and the pharynx.The animal is, however, quite capable of functioning and moving about if the nerve ring is surgically removed, demonstrating that it does not have a central role in nervous coordination. In addition, five major nerves run from the nerve ring down the length of the body beneath each of the ambulacral areas.

Most sea cucumbers have no distinct sensory organs, although there are various nerve endings scattered through the skin giving the animal a sense of touch and a sensitivity to the presence of light. There are, however, a few exceptions; members of the Apodida order are known to possess statocysts, while some species possess small eye-spots near the bases of their tentacles.

Respiratory System

Sea cucumbers extract oxygen from water in a pair of "respiratory trees" that branch off the cloaca just inside the anus, so that they "breathe" by drawing water in through the anus and then expelling it. The trees consist of a series of narrow tubules branching from a common duct, and lie on either side of the digestive tract. Gas exchange occurs across the thin walls of the tubules, to and from the fluid of the main body cavity.

Together with the intestine, the respiratory trees also act as excretory organs, with nitrogenous waste diffusing across the tubule walls in the form of ammonia and phagocytic coelomocytes depositing particulate waste.

Circulatory Systems

Like all echinoderms, sea cucumbers possess both a water vascular system that provides hydraulic pressure to the tentacles and tube feet, allowing them to move, and a *haemal system*. The latter is more complex than that in other echinoderms, and consists of well-developed vessels as well as open sinuses.

A central haemal ring surrounds the pharynx next to the ring canal of the water vascular system, and sends off additional vessels along the radial canals beneath the ambulacral areas. In the larger species, additional vessels run above and below the intestine and are connected by over a hundred small muscular ampullae, acting as miniature hearts to pump blood around the haemal system. Additional vessels surround the respiratory trees, although they contact them only indirectly, via the coelomic fluid.

Indeed, the blood itself is essentially identical with the coelomic fluid that bathes the organs directly, and also fills the water vascular system. Phagocytic coelomocytes, somewhat similar in function to the white blood cells of vertebrates, are formed within the haemal vessels, and travel throughout the body cavity as well as both circulatory systems. An additional form of coelomocyte, not found in other echinoderms, has a flattened discoid shape, and contains hemoglobin. As a result, in many (though not all) species, both the blood and the coelomic fluid are red in colour.

Pearsonothuria graeffei showing its three row of podia on its *trivium*

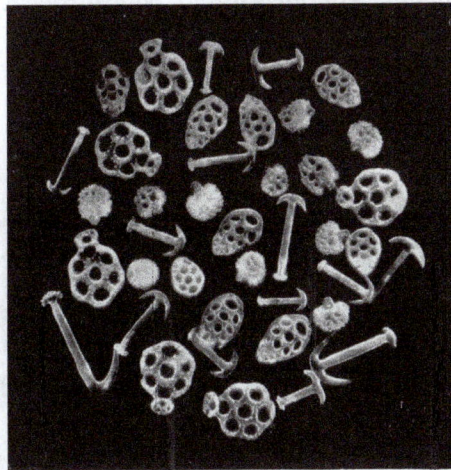

Sea cucumber ossicles (here "wheels" and "anchors")

Vanadium has been reported in high concentrations in holothurian blood, however researchers have been unable to reproduce these results.

Locomotive Organs

Like all echinoderms, sea cucumbers possess pentaradial symmetry. However, because of their posture, they have secondarily evolved a degree of bilateral symmetry. For example, because one side of the body is typically pressed against the substratum, and the other is not, there is usually some difference between the two surfaces (except for Apodida). Like sea urchins, most sea cucumbers have five strip-like ambulacral areas running along the length of the body from the mouth to the anus. The three on the lower surface have numerous tube feet, often with suckers, that allow the animal to crawl along; they are called *trivium*. The two on the upper surface have under-developed or vestigial tube feet, and some species lack tube feet altogether; this face is called *bivium*.

In some species, the ambulacral areas can no longer be distinguished, with tube feet spread over a much wider area of the body. Those of the order Apodida have no tube feet or ambulacral areas at all, and burrow through sediment with muscular contractions of their body similar to that of worms, however five radial lines are generally still obvious along their body.

Even in those sea cucumbers that lack regular tube feet, those that are immediately around the mouth are always present. These are highly modified into retractile tentacles, much larger than the locomotive tube feet. Depending on the species, sea cucumbers have between ten and thirty such tentacles and these can have a wide variety of shapes depending on the diet of the animal and other conditions.

Many sea cucumbers have papillae, conical fleshy projections of the body wall with sensory tube feet at their apices. These can even evolve into long antennae-like structures, especially on the abyssal genus *Scotoplanes*.

Endoskeleton

Echinoderms typically possess an internal skeleton composed of plates of calcium carbonate. In most sea cucumbers, however, these have become reduced to microscopic ossicles embedded beneath the skin. A few genera, such as *Sphaerothuria*, retain relatively large plates, giving them a scaly armour.

Life History and Behaviour

Habitat

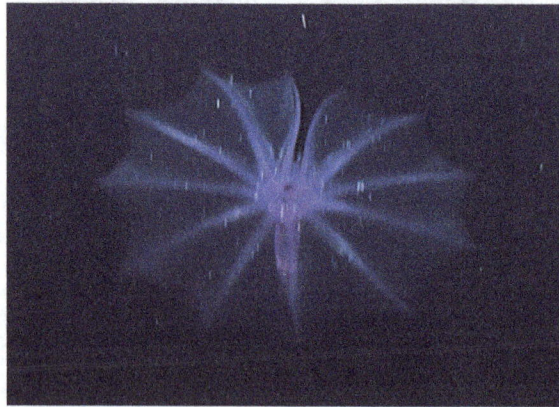

The mysterious *Pelagothuria natatrix* is the only truly pelagic echinoderm known to date.

Sea cucumbers can be found in great numbers on the deep seafloor, where they often make up the majority of the animal biomass. At depths deeper than 8.9 km (5.5 mi), sea cucumbers comprise 90% of the total mass of the macrofauna. Sea cucumbers form large herds that move across the bathygraphic features of the ocean, hunting food. The body of some deep water holothurians, such as *Enypniastes eximia*, *Peniagone leander* and *Paelopatides confundens*, is made of a tough gelatinous tissue with unique properties that makes the animals able to control their own buoyancy, making it possible for them to either live on the ocean floor or to actively swim or float over it in order to move to new locations, in a manner similar to how the group Torquaratoridae floats through water.

Holothurians appear to be the best adapted echinoderms to extreme depths, and are still very diversified beyond 5,000 m deep: several species from the Elpidiidae family ("sea pigs") can be found deeper than 9,500 m, and the record seems to be some species of the *Myriotrochus* genus (in particular *Myriotrochus bruuni*), identified down to 10,687 meters deep. In more shallow waters, sea cucumbers can form dense populations. The strawberry sea cucumber (*Squamocnus brevidentis*) of New Zealand lives on rocky walls around the southern coast of the South Island where populations sometimes reach densities of 1,000 animals per square meter (93 /sq ft). For this reason, one such area in Fiordland is called the strawberry fields.

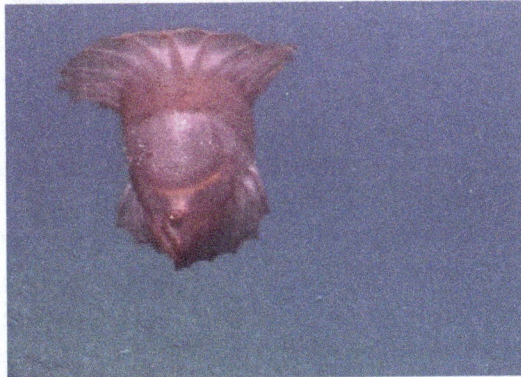

Benthopelagic sea cucumbers, such as this *Enypniastes*, are often confused
with jellyfish, have webbed swimming structures enabling them to swim up off the surface of the seafloor
and journey as much as 1,000 m (3,300 ft) up the water column

Spanish dancer (*Benthodytes* sp.), another swimming sea cucumber,
hovering at 2789 meters by the Davidson Seamount

Locomotion

Some abyssal species in the abyssal order Elasipodida have evolved to a "benthopelagic" behaviour: their body is nearly the same density as the water around them, so they can make long jumps (up to 1,000 m (3,300 ft) high), before falling slowly back to the ocean floor. Most of them have specific swimming appendages, such as some kind of umbrella (like *Enypniastes*), or a long lobe on top of the body (*Psychropotes*). Only one species is known as a true completely pelagic species, that never comes close to the bottom: *Pelagothuria natatrix*.

Diet

Holothuroidea are generally scavengers, feeding on debris in the benthic zone of the ocean. Exceptions include some pelagic cucumbers and the species *Rynkatorpa pawsoni*, which has a commensal relationship with deep-sea anglerfish. The diet of most cucumbers consists of plankton and decaying organic matter found in the sea. Some sea cucumbers position themselves in currents and catch food that flows by with their open tentacles. They also sift through the bottom sediments using their tentacles. Other species can dig into bottom silt or sand until they are completely buried. They then extrude their feeding tentacles, ready to withdraw at any hint of danger.

In the South Pacific sea cucumbers may be found in densities of 40 individuals per square meter (33 /sq yd). These populations can process 19 kilograms of sediment per square meter (34 lb /sq yd) per year.

The shape of the tentacles is generally adapted to the diet, and to the size of the particles to be ingested: the filter-feeding species mostly have complex arborescent tentacles, intended to maximize the surface area available for filtering, while the species feeding on the substratum will more often need digitate tentacles to sort out the nutritional material; the detritivore species living on fine sand or mud more often need shorter "peltate" tentacles, shaped like shovels. A single specimen can swallow more than 45 kg of sediment a year, and their excellent digestive capacities allow them to reject a finer, purer and homogeneous sediment. Therefore, sea cucumbers play a major role in the biological processing of the sea bed (bioturbation, purge, homogenization of the substratum etc.).

| The mouth of an *Euapta godeffroyi*, showing pinnate tentacles. | Mouth of a *Holothuria sp.*, showing peltate tentacles. | Mouth of a *Cucumaria miniata*, with dendritic tentacles, for filtering the water. | Faeces of an holothurian. This participates in sediment homogenization and purification. |

Communication and Sociability

Sea cucumbers communicate with each other by sending hormone signals through the water. The main aim of communication is reproduction; otherwise they tend to ignore each other. They do not show territorial behaviour. Some species, including abyssal species such as *Scotoplanes globosa*, can live in groups.

Reproduction

Most sea cucumbers reproduce by releasing sperm and ova into the ocean water. Depending on conditions, one organism can produce thousands of gametes. Sea cucumbers are typically dioecious, with separate male and female individuals, but some species are protandric. The reproductive system consists of a single gonad, consisting of a cluster of tubules emptying into a single duct that opens on the upper surface of the animal, close to the tentacles.

At least 30 species, including the red-chested sea cucumber (*Pseudocnella insolens*), fertilize their eggs internally and then pick up the fertilized zygote with one of their feeding tentacles. The egg is then inserted into a pouch on the adult's body, where it develops and eventually hatches from the pouch as a juvenile sea cucumber. A few species are known to brood their young inside the body cavity, giving birth through a small rupture in the body wall close to the anus.

"Auricularia" larva(by Ernst Haeckel).

Development

In all other species, the egg develops into a free-swimming larva, typically after around three days of development. The first stage of larval development is known as an auricularia, and is only around 1 mm (39 mils) in length. This larva swims by means of a long band of cilia wrapped around its body, and somewhat resembles the bipinnaria larva of starfish. As the larva grows it transforms into the doliolaria, with a barrel-shaped body and three to five separate rings of cilia. The "pentacularia" is the third larval stage of sea cucumber, where the tentacles appear. The tentacles are usually the first adult features to appear, before the regular tube feet.

Symbiosis and Commensalism

Emperor shrimp *Periclimenes imperator* on a *Bohadschia ocellata* sea cucumber.

Numerous small animals can live in symbiosis or commensalism with sea cucumbers, as well as some parasites.

Some cleaner shrimps can live on the tegument of holothurians, in particular several species of the genus *Periclimenes* (genus which is specialized in echinoderms), in particular *Periclimenes imperator*. A variety of fish, most commonly pearl fish, have evolved a commensalistic symbiotic relationship with sea cucumbers in which the pearl fish will live in sea cucumber's cloaca using it for protection from predation, a source of food (the nutrients passing in and out of the anus from the water), and to develop into their adult stage of life. Many polychaete worms (family Polynoidae) and crabs (like *Lissocarcinus orbicularis*) have also specialized to use the mouth or the cloacal respiratory trees for protection by living inside the sea cucumber. Nevertheless, holothurians species of the genus *Actinopyga* have anal teeth that prevent visitors from penetrating their anus.

Sea cucumbers can also shelter bivalvia as endocommensals, such as *Entovalva sp.*

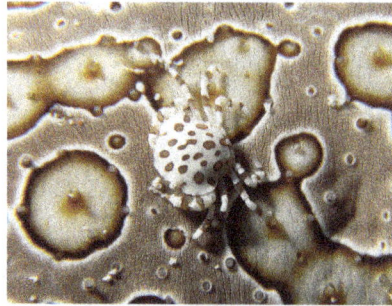

Lissocarcinus orbicularis, a symbiotic crab.

Periclimenes imperator, a symbiotic shrimp.

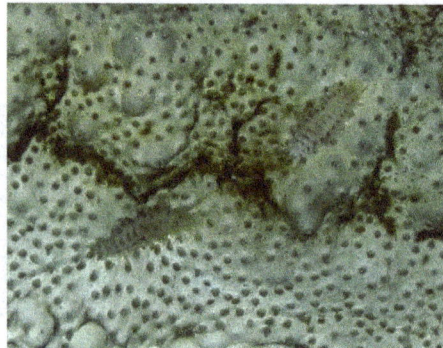

Polynoid worms on a king sea cucumber.

Predators and Defensive Systems

Sea cucumbers are often scorned by most of the marine predators because of the toxins they contain (in particular holothurin) and because of their often spectacular defensive systems. However, they remain a prey for some highly specialized predators which are not affected by their toxins, such as the big mollusks *Tonna galea* and *Tonna perdix*, which paralyzes them using powerful poison before swallowing them completely. Some other less specialized and opportunist predators can also prey on sea cucumbers sometimes when they can't find any better food, such as certain species of fish (triggerfish, pufferfish) and crustaceans (crabs, lobsters, hermit crabs).

Tonna perdix, a selective predator of tropical sea cucumbers.

A sea cucumber in Mahé, Seychelles ejects sticky filaments from the anus in self-defense.

Some species of coral-reef sea cucumbers within the order Aspidochirotida can defend themselves by expelling their sticky cuvierian tubules (enlargements of the respiratory tree that float freely in the coelom) to entangle potential predators. When startled, these cucumbers may expel some of them through a tear in the wall of the cloaca in an autotomic process known as evisceration. Replacement tubules grow back in one and a half to five weeks, depending on the species. The release of these tubules can also be accompanied by the discharge of a toxic chemical known as holothurin, which has similar properties to soap. This chemical can kill animals in the vicinity and is one more method by which these sedentary animals can defend themselves.

Estivation

If the water temperature becomes too high, some species of sea cucumber from temperate seas can aestivate. While they are in this state of dormancy they stop feeding, their gut atrophies, their metabolism slows down and they lose weight. The body returns to its normal state when conditions improve.

Phylogeny and Classification

Holothuries having no skeleton unlike the other echinoderms, their classification is more complex and their paleontological phylogeny relies on a fistful of well-preserved specimens. The modern taxonomy is based first of all on the presence or the shape of certain soft parts (podia, lungs, tentacles, peripharingal crown) to determine the main orders, and secondarily on the microscopic examination of ossicles to determine the genus and the species. The contemporary genetic methods largely helped to make the classification of these animals progress.

Apodida like this *Euapta godeffroyi* are snake-shaped, without podia, and have pinnate tentacles.

Aspidochirotida like this *Holothuria cinerascens* are sausage-shaped, with peltate tentacles.

Dendrochirotida like this *Cercodemas anceps* are curled-bodied and have arborescent tentacles.

Taxonomic classification according to World Register of Marine Species :

- order Apodida *Brandt, 1835*
- family Chiridotidae Östergren, 1898
- family Myriotrochidae Théel, 1877
- family Synaptidae Burmeister, 1837

- order Aspidochirotida Grube, 1840
- family Holothuriidae Burmeister,1837
- family Mesothuriidae Smirnov,2012
- family Stichopodidae Haeckel,1896
- family Synallactidae Ludwig, 1894
- order Dendrochirotida Grube, 1840
- family Cucumariidae Ludwig, 1894
- family Cucumellidae Thandar & Arumugam,2011
- family Heterothyonidae Pawson,1970
- family Paracucumidae Pawson & Fell,1965
- family Phyllophoridae Östergren,1907
- family Placothuriidae Pawson & Fell,1965
- family Psolidae Burmeister, 1837
- family Rhopalodinidae Théel, 1886
- family Sclerodactylidae Panning,1949
- family Vaneyellidae Pawson & Fell,1965
- family Ypsilothuriidae Heding,1942
- order Elasipodida Théel, 1882
- family Deimatidae Théel, 1882
- family Elpidiidae Théel, 1882
- family Laetmogonidae Ekman, 1926
- family Pelagothuriidae Ludwig,1893
- family Psychropotidae Théel, 1882
- order Molpadida Haeckel, 1896
- family Caudinidae Heding, 1931
- family Eupyrgidae Semper, 1867
- family Gephyrothuriidae Koehler & Vaney,1905
- family Molpadiidae Müller, 1850

Holothurians as Food And Medicine

Dried sea cucumbers in a Japanese pharmacy

To supply the markets of Southern China, Makassar trepangers traded with the Indigenous Australians of Arnhem Land. This is the first recorded example of trade between the inhabitants of the Australian continent and their Asian neighbours.

There are many commercially important species of sea cucumber that are harvested and dried for export for use in Chinese cuisine as *hoisam*. Some of the more commonly found species in markets include:

- *Holothuria nobilis*
- *Thelenota ananas*
- *Actinopyga echinites*
- *Actinopyga palauensis*
- *Holothuria scabra*
- *Holothuria fuscogilva*
- *Actinopyga mauritiana*
- *Stichius japonicus*
- *Parastichopus californicus*
- *Acaudina molpadioides*
- *Isostichopus fuscus*

According to the American Cancer Society, although it has been used in traditional Asian folk medicine for a variety of ailments, "there is little reliable scientific evidence to support claims that sea cucumber is effective in treating cancer, arthritis, and other diseases" but there is research being conducted to determine "whether some compounds made by sea cucumbers may be helpful against cancer."

There are pharmaceutical companies being built based on *gamat*, the traditional medicinal usage of this animal. Extracts are prepared and made into oil, cream, or cosmetics. Some products are intended to be taken internally. A single study conducted on an unreported number of mice found intraperitoneal injection of sea cucumber extract to be somewhat effective in high doses (100 ppm or mg/kg) against internal pain, but ineffective against externally induced pain. Another study suggested that the sea cucumber contains all the fatty acids necessary to play a potentially active role in tissue repair. Sea cucumbers are being researched for use in treating a wide number of ailments; including colon cancer. Surgical probes made of nanocomposite material based off the sea cucumber have been shown to reduce brain scarring. One study found that a lectin from *Cucumaria echinata* impaired the development of the malaria parasite when produced by transgenic mosquitoes.

| "Teripang" in a market in Asia. | Sea cucumber in sauce in Chin | *Haisom cah jamur*, Chinese Indonesian sea cucumber with mushroom. | Deep fried sea cucumbers. |

Commercial Harvest

In recent years, the sea cucumber industry in Alaska has increased due to increased export of the skins and muscles to China.

In China, sea cucumbers are farmed commercially in artificial ponds. These ponds can be as large as 400 hectares (1,000 acres) and satisfy much of the local demand. Wild sea cucumbers are caught by divers and these wild Alaskan sea cucumbers have higher nutritional value and are larger than farmed Chinese sea cucumbers. Larger size and higher nutritional value has allowed the Alaskan fisheries to continue to compete for market share, despite the increase in local, Chinese sea cucumber farming.

One of Australia's oldest fisheries is the collection of sea cucumber, harvested by divers from throughout the Coral Sea in far North Queensland, Torres Straits and Western Australia. In the late 1800s there were as many as 400 divers operating from Cook Town, Queensland.

Overfishing of sea cucumbers in the Great Barrier Reef is threatening their population. Their popularity as luxury seafood in the Asian countries poses a serious threat to the order Aspidochirotida.

Black Market

As of 2013 there was a thriving black market in sea cucumbers driven by demand in China where 1 lb (0.5 kg) at its peak might have sold for the equivalent of US$300 and a single sea cucumber for about US$160. However, a crackdown by governments both in and out of China reduced both prices and consumption, particularly among government officials who had been known to eat (and were able to afford purchasing) the most expensive and rare species. In the Caribbean Sea off the shores of the Yucatán Peninsula near fishing ports such as Dzilam de Bravo, illegal harvesting had devastated the population of sea cucumbers and resulted in conflict in the community as rival gangs struggled to control the illegal harvest. They are normally dried for selling.

Aquaculture

Overexploitation of sea cucumber stocks in many parts of the world provided motivation for the development of sea cucumber aquaculture in the early 1980s. The Chinese and Japanese were the first to develop successful hatchery technology on *Apostichopus japonicus*, prized for its high meat content and success in commercial hatcheries. Using the techniques pioneered by the Chinese and Japanese, a second species, *Holothuria scabra*, was cultured for the first time in India in 1988. In recent years Australia, Indonesia, New Caledonia, Maldives, Solomon Islands and Vietnam have also successfully cultured *H. scabra* using the same technology, which has since been expanded to other species.

In Art and Literature

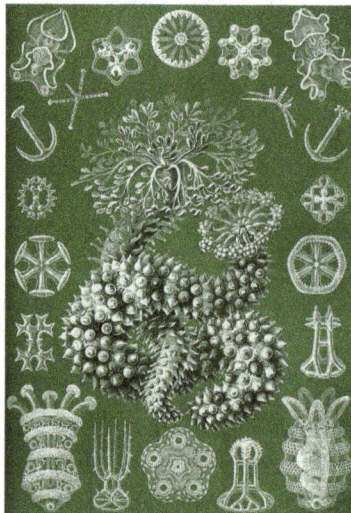

Holothurians plate by Ernst Haeckel from his *Kunstformen der Natur* (1904).

Edgar Allan Poe's only novel, *The Narrative of Arthur Gordon Pym of Nantucket* (1838), includes in its twentieth chapter a long, detailed description of sea cucumbers, which the narrator calls *biche de mer*.

The first movement of French pianist Erik Satie's Embryons desséchés is titled "D'Holothournie". It is said to emulate the "purring" of the Holothourian.

Sea cucumbers have inspired thousands of haiku in Japan, where they are called *namako* written with characters that can be translated as "sea mice" (an example of gikun). In English translations of these haiku, they are usually called "sea slugs». According to the *Oxford English Dictionary*, the English term "sea slug" was originally applied to holothurians during the 18th century. The term is now applied to several groups of sea snails, marine gastropod mollusks that have no shell or only a very reduced shell, including the nudibranchs. Almost 1,000 Japanese holothurian haiku translated into English appear in the book *Rise, Ye Sea Slugs!* by Robin D. Gill.

Nobel laureate poet Wisława Szymborska wrote a poem which mentions holothurians, titled "Autotomy".

In the book *John Dies at the End*, the character Amy Sullivan was nicknamed "Cucumber" by the narrator/author when the two were children. This is assumed by other characters to have a sexual connotation but is actually a reference to her frequent nausea. The name is in reference to a sea cucumber's use of vomiting as a method of self-defense.

A descriptive passage in American novelist and MacArthur Fellow Cormac McCarthy's 1985 anti-Western *Blood Meridian* likens cactus embers to holothurians: "In the thorn forest through which they'd passed the little desert wolves yapped and on the dry plain before them others answered and the wind fanned the coals that he watched. The bones of cholla that glowed there in their incandescent basketry pulsed like burning holothurians in the phosphorous dark of the sea's deeps."

In more recent times, the web comic Jerkcity has incorporated references to sea cucumbers.

Crinoid

Crinoids are marine animals that make up the class Crinoidea of the echinoderms (phylum Echinodermata). The name comes from the Greek word *krinon*, "a lily", and *eidos*, "form". They live in both shallow water and in depths as great as 9,000 meters (30,000 ft). Those crinoids which in their adult form are attached to the sea bottom by a stalk are commonly called sea lilies. The unstalked forms are called feather stars or comatulids.

Crinoids are characterised by a mouth on the top surface that is surrounded by feeding arms. They have a U-shaped gut, and their anus is located next to the mouth. Although the basic echinoderm pattern of fivefold symmetry can be recognised, most crinoids have many more than five arms. Crinoids usually have a stem used to attach themselves to a substrate, but many live attached only as juveniles and become free-swimming as adults.

There are only about 600 extant crinoid species, but they were much more abundant and diverse in the past. Some thick limestone beds dating to the mid- to late-Paleozoic are almost entirely made up of disarticulated crinoid fragments.

Morphology

A stalked crinoid drawn by Ernst Haeckel

Crinoids comprise three basic sections; the stem, the calyx, and the arms. The stem is composed of highly porous ossicles which are connected by ligamentary tissue. The calyx contains the crinoid's digestive and reproductive organs, and the mouth is located at the top of the dorsal cup, while the anus is located peripheral to it. The arms display pentamerism or pentaradial symmetry and comprise smaller ossicles than the stem and are equipped with cilia which facilitate feeding by moving the organic media down the arm and into the mouth.

Close-up on the calyx of a characteristic abyssal stalked crinoid.
Ten arms are visible, with their pinnules.

The majority of living crinoids are free-swimming and have only a vestigial stalk. In those deep-sea species that still retain a stalk, it may reach up to 1 metre (3.3 ft) in length, although it is usually much smaller. The stalk grows out of the *aboral* surface, which forms the upper side of the animal in starfish and sea urchins, so that crinoids are effectively upside-down by comparison with most other echinoderms. The base of the stalk consists of a disc-like sucker, which, in some species, has root-like structures that further increase its grip on the underlying surface. The stalk is often lined by small cirri.

Like other echinoderms, crinoids have pentaradial symmetry. The aboral surface of the body is studded with plates of calcium carbonate, forming an endoskeleton similar to that in starfish and sea urchins. These make the calyx somewhat cup-shaped, and there are few, if any, ossicles in the oral (upper) surface. The upper surface, or *tegmen*, is divided into five *ambulacral areas*, including a deep groove from which the tube feet project, and five *interambulacral areas* between them. The anus, unusually for echinoderms, is found on the same surface as the mouth, at the edge of the tegmen.

Stem, theca and arms of a "true" (stalked) crinoid (family Isselicrinidae).

The ambulacral grooves extend onto the arms, which thus have tube feet along their inner surfaces.

Primitively, crinoids had only five arms, but in most living species these are divided into two, giving ten arms in total. In most living species, especially the free-swimming feather stars, the arms branch several times, producing up to two hundred branches in total. The arms are jointed, and lined by smaller feather-like appendages, or *pinnules*, which also include tube feet.

Oxycomanthus bennetti (comatulid)

Tegmen of a *Lamprometra palmata*. The mouth is located at the center of the 5 feeding grooves, and the anus at the top of the column.

Biology

Feeding

Close-up on the pinnules with visible rows of translucent podia.

The mouth descends into a short oesophagus. There is no true stomach, so the oesophagus connects directly to the intestine, which runs in a single loop right around the inside of the calyx. The intestine often includes numerous diverticulae, some of which may be long or branched. The end of the intestine opens into a short muscular rectum. This ascends towards the anus, which projects from a small conical protuberance at the edge of the tegmen.

Predation

Specimens of the sea urchin *Calocidaris micans* present in a meadow of the crinoid *Endoxocrinus parrae*, have been shown to contain large quantities of stem portions (or columnals) in the direct vicinity of living crinoids, some of which were upended. The gut content of the sea urchins consisted of articulated ossicles with soft tissue, whereas the local sediment contained only disarticulated ossicles without soft tissue. This makes it highly likely that these sea urchins are predators of the crinoids, and that the crinoids flee, offering part of their stem in the process.

Various crinoid fossils hint at possible prehistoric predators. Coprolites of both fish and cephalopods have been found containing ossicles of various crinoids, such as the pelagic crinoid *Saccocoma*, from the Jurassic lagerstatten Solnhofen, while damaged crinoid stems with bite marks matching the toothplates of coccosteid placoderms have been found in Late Devonian Poland. The calyxes of several Devonian to Carboniferous-aged crinoids have the shells of a snail, *Platyceras*, intimately associated with them. Some have the snail situated over the anus, suggesting that *Platyceras* was a coprophagous commensal, while others have the animal directly situated over a borehole, suggesting a more pernicious relationship.

Circulatory Systems

Like other echinoderms, crinoids possess a water vascular system that maintains hydraulic pressure in the tube feet. This is not connected to external sea water, as in other echinoderms, but only to the body cavity. The body cavity is itself somewhat restricted, being largely replaced by connective tissue, although it is present as narrow canals within the arms and stalk.

Crinoids also possess a separate haemal system, consisting of fluid-filled sinuses within the connective tissue. There is a large plexus of sinuses around the oesophagus, with branches extending down to a mass of glandular tissue at the base of the calyx.

These various fluid-filled spaces, in addition to transporting nutrients around the body, also function as both a respiratory and an excretory system. Oxygen is absorbed primarily through the tube feet, which are the most thin-walled parts of the body, while waste is collected by phagocytic coelomocytes.

Nervous System

The crinoid nervous system is divided into three parts, with numerous connections between them. The uppermost portion is the only one homologous with the nervous systems of other echinoderms. It consists of a central nerve ring surrounding the mouth, and radial nerves branching into the arms. Below this lies a second nerve ring, giving off two brachial nerves into each arm. Both of these sets of nerves are sensory in nature, with the lower set supplying the pinnules and tube feet. The third portion of the nervous system lies below the other two, and is responsible for motor action. This is centred on a mass of neural tissue near the base of the calyx, and provides a single nerve to each arm and a number of nerves to the stalk.

Reproduction and Life Cycle

Crinoids are dioecious, with separate male and female individuals. They have no true gonads, producing their gametes from genital canals found inside some of the pinnules. The pinnules eventually rupture to release the sperm and eggs into the surrounding sea water. The fertilised eggs hatch to release a free-swimming vitellaria larva. The larva is barrel-shaped with rings of cilia running round the body, and a tuft of sensory hairs at the upper pole. In some cases females have been known to temporarily brood the larvae using chambers within the arms. While both feeding and non-feeding larvae exist among the four other extant echinoderm classes, all present day crinoids appears to be descendants from a surviving clade that went through a bottleneck after the Permian extinction, which had lost its feeding larval stage. The larva's free-swimming period lasts only for a few days before settling to the bottom and attaching itself to the underlying surface using an adhesive gland on its ventral surface. The larva then metamorphoses into a stalked juvenile. Even the free-swimming feather stars sometimes go through this stage, with the adult eventually breaking away from the stalk. Within 10 to 16 months the crinoid will be able to reproduce.

Mobility

A stalked crinoid (white) and a comatulid (red) in deep sea, showing the differences between these two sister groups.

Most modern crinoids, i.e., the feather stars, are free-swimming and lack a stem as adults. Examples of fossil crinoids that have been interpreted as free-swimming include *Marsupitsa*, *Saccocoma* and *Uintacrinus*. In 2005, a stalked crinoid was recorded pulling itself along the sea floor off the Grand Bahama Island. While it has been known that stalked crinoids move, before this recording the fastest motion of a crinoid was 0.6 metres/hour (2 ft/h) 0.167 mm/s (millimeters per second). The 2005 recording showed a crinoid moving much faster, at a rate of 4-5 centimeters/second (144 to 180 meters per hour).

Evolution

If one ignores the enigmatic *Echmatocrinus* of the Burgess Shale, the earliest known unequivocal crinoid groups date back to the Ordovician. There are two competing hypotheses pertaining to the origin of the group: the traditional viewpoint holds that crinoids evolved from within the blastozoans (the eocrinoids and their derived descendants the cystoids), whereas the most popular alternative suggests that the crinoids split early from among the edrioasteroids. The debate is difficult to settle, in part because all three candidate ancestors share many characteristics, including radial symmetry, calcareous plates, and stalked or direct attachment to the substrate.

Agaricocrinus americanus, a fossil crinoid from the Carboniferous of Indiana.

Middle Jurassic (Callovian) *Apiocrinites* crinoid pluricolumnals from the Matmor Formation in Hamakhtesh Hagadol, southern Israel.

Diversity

The crinoids underwent two periods of abrupt adaptive radiation; the first during the Ordovician, the other was during the early Triassic after they underwent a selective mass extinction at the end of the Permian period. This Triassic radiation resulted in forms possessing flexible arms becoming widespread; motility, predominantly a response to predation pressure, also became far more prevalent. This radiation occurred somewhat earlier than the Mesozoic marine revolution, possibly because it was mainly prompted by increases in benthic predation, specifically of echinoids. After the end-Permian extinction, crinoids never regained the morphological diversity they enjoyed in the Paleozoic; they employed a different suite of the ecological strategies open to them from those that had proven so successful in the Paleozoic.

The long and varied geological history of the crinoids demonstrates how well the echinoderms have adapted to filter-feeding. The fossils of other stalked filter-feeding echinoderms, such as blastoids, are also found in rocks of the Palaeozoic era. These extinct groups can exceed the crinoids in both numbers and variety in certain strata. However, none of these others survived the crisis at the end of the Permian period.

Fossils of Interest

Some fossil crinoids, such as *Pentacrinites*, seem to have lived attached to floating driftwood and complete colonies are often found. Sometimes this driftwood would become waterlogged and sink to the bottom, taking the attached crinoids with it. The stem of *Pentacrinites* can be several metres long. Modern relatives of *Pentacrinites* live in gentle currents attached to rocks by the end of their stem. The largest fossil crinoid on record had a stem 40 m (130 ft) in length.

In 2012, three geologists reported they had isolated complex organic molecules from 340-million-year-old (Mississippian) fossils of multiple species of crinoids. Identified as "resembl[ing ...] aromatic or polyaromatic quinones", these are the oldest molecules to be definitively associated with particular individual fossils, as they are believed to have been sealed inside ossicle pores by precipitated calcite during the fossilization process.

Crinoid Uses

- Fossilised crinoid columnal segments extracted from limestone quarried on Lindisfarne, or found washed up along the foreshore, were threaded into necklaces or rosaries, and became known as St. Cuthbert's beads.

- In the Midwestern United States, fossilized segments of columnal crinoids are sometimes known as Indian beads.

- Crinoids are the state fossil of Missouri.

Cnidaria

Cnidaria is a phylum containing over 10,000 species of animals found exclusively in aquatic (freshwater and marine) environments: they are predominantly marine species. Their distinguishing feature is cnidocytes, specialized cells that they use mainly for capturing prey. Their bodies consist of mesoglea, a non-living jelly-like substance, sandwiched between two layers of epithelium that are mostly one cell thick. They have two basic body forms: swimming medusae and sessile polyps, both of which are radially symmetrical with mouths surrounded by tentacles that bear cnidocytes. Both forms have a single orifice and body cavity that are used for digestion and respiration. Many cnidarian species produce colonies that are single organisms composed of medusa-like or polyp-like zooids, or both (hence they are trimorphic). Cnidarians' activities are coordinated by a decentralized nerve net and simple receptors. Several free-swimming species of Cubozoa and Scyphozoa possess balance-sensing statocysts, and some have simple eyes. Not all cnidarians reproduce sexually, with many species having complex life cycles of asexual polyp stages and sexual medusae. Some, however, omit either the polyp or the medusa stage.

Cnidarians were formerly grouped with ctenophores in the phylum Coelenterata, but increasing awareness of their differences caused them to be placed in separate phyla. Cnidarians are classified into four main groups: the almost wholly sessile Anthozoa (sea anemones, corals, sea pens); swimming Scyphozoa (jellyfish); Cubozoa (box jellies); and Hydrozoa, a diverse group that includes all the freshwater cnidarians as well as many marine forms, and has both sessile members, such as *Hydra*, and colonial swimmers, such as the Portuguese Man o' War. Staurozoa have recently been recognised as a class in their own right rather than a sub-group of Scyphozoa, and the parasitic Myxozoa and Polypodiozoa were only firmly recognized as cnidarians in 2007.

Most cnidarians prey on organisms ranging in size from plankton to animals several times larger than themselves, but many obtain much of their nutrition from dinoflagellates, and a few are parasites. Many are preyed on by other animals including starfish, sea slugs, fish and turtles. Many scleractinian corals—which form the structural foundation for coral reefs—possess polyps that are filled with symbiotic photo-synthetic zooxanthellae. While reef-forming corals are almost entirely restricted to warm and shallow marine waters, other cnidarians can be found at great depths, in polar regions, and in freshwater.

Recent phylogenetic analyses support monophyly of cnidarians, as well as the position of cnidarians as the sister group of bilaterians. Fossil cnidarians have been found in rocks formed about 580 million years ago, and other fossils show that corals may have been present shortly before 490 million years ago and diversified a few million years later. However, molecular clock analysis of mitochondrial genes suggests a much older age for the crown group of cnidarians, estimated around 741 million years ago, almost 200 million years before the Cambrian period as well as any fossils.

Distinguishing Features

Cnidarians form an animal phylum that are more complex than sponges, about as complex as ctenophores (comb jellies), and less complex than bilaterians, which include almost all other animals. However, both cnidarians and ctenophores are more complex than sponges as they have: cells bound by inter-cell connections and carpet-like basement membranes; muscles; nervous systems; and some have sensory organs. Cnidarians are distinguished from all other animals by having cnidocytes that fire like harpoons and are used mainly to capture prey. In some species, cnidocytes can also be used as anchors.

Like sponges and ctenophores, cnidarians have two main layers of cells that sandwich a middle layer of jelly-like material, which is called the mesoglea in cnidarians; more complex animals have three main cell layers and no intermediate jelly-like layer. Hence, cnidarians and ctenophores have traditionally been labelled diploblastic, along with sponges. However, both cnidarians and ctenophores have a type of muscle that, in more complex animals, arises from the middle cell layer. As a result, some recent text books classify ctenophores as triploblastic, and it has been suggested that cnidarians evolved from triploblastic ancestors.

	Sponges	Cnidarians	Ctenophores	Bilateria
Cnidocytes	No	Yes	No	
Colloblasts	No		Yes	No
Digestive and circulatory organs	No			Yes

Number of main cell layers	Two, with jelly-like layer between them		Two or Three	Three
Cells in each layer bound together	cell-adhesion molecules, but no basement membranes except Homoscleromorpha.	inter-cell connections; basement membranes		
Sensory organs	No	Yes		
Number of cells in middle "jelly" layer	Many	Few		(Not applicable)
Cells in outer layers can move inwards and change functions	Yes	No		(Not applicable)
Nervous system	No	Yes, simple		Simple to complex
Muscles	None	Mostly epitheliomuscular	Mostly myoepithelial	Mostly myocytes

Description

Basic Body Forms

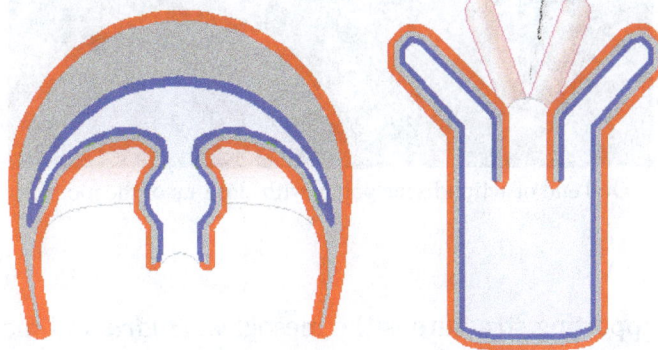

Medusa (left) and polyp (right)

Exoderm   Gastroderm (Endoderm)   Mesoglea   Digestive cavity

Most adult cnidarians appear as either swimming medusae or sessile polyps, and many hydrozoan species are known to alternate between the two forms. Both are radially symmetrical , like a wheel and a tube respectively. Since these animals have no heads, their ends are described as "oral" (nearest the mouth) and "aboral" (furthest from the mouth). Most have fringes of tentacles equipped with cnidocytes around their edges, and medusae generally have an inner ring of tentacles around the mouth. Some hydroids may consist of colonies of zooids that serve different purposes, such as defense, reproduction and catching prey. The mesoglea of polyps is usually thin and often soft, but that of medusae is usually thick and springy, so that it returns to its original shape after muscles around the edge have contracted to squeeze water out, enabling medusae to swim by a sort of jet propulsion.

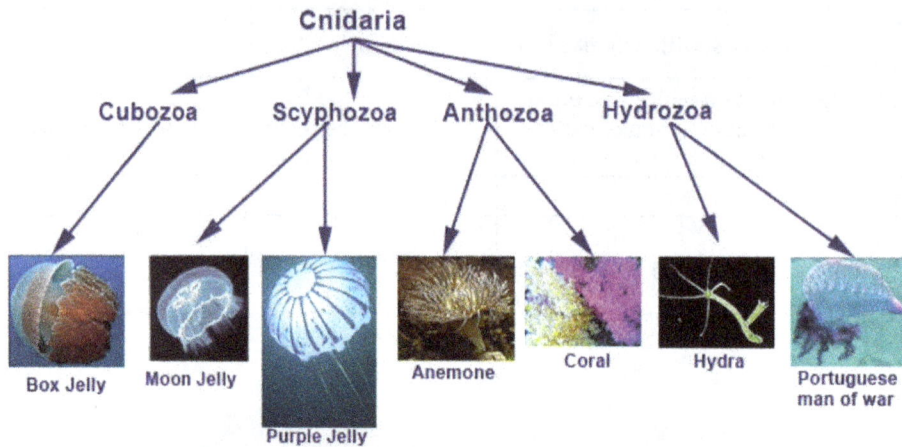

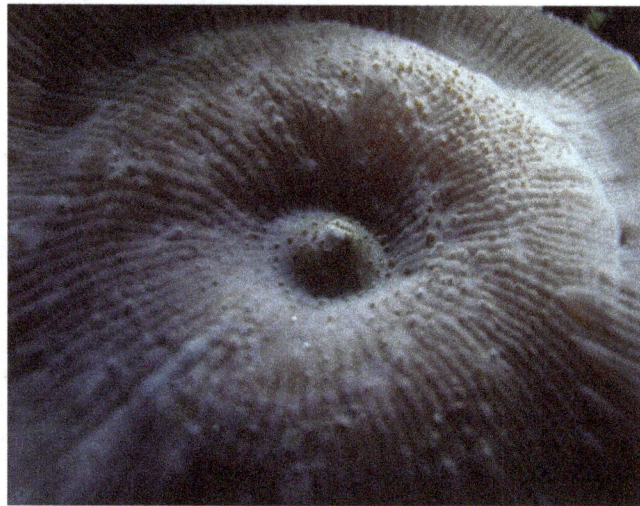

Oral end of actinodiscus polyp, with close-up of the mouth

Skeletons

In medusae the only supporting structure is the mesoglea. *Hydra* and most sea anemones close their mouths when they are not feeding, and the water in the digestive cavity then acts as a hydrostatic skeleton, rather like a water-filled balloon. Other polyps such as *Tubularia* use columns of water-filled cells for support. Sea pens stiffen the mesoglea with calcium carbonate spicules and tough fibrous proteins, rather like sponges.

In some colonial polyps, a chitinous periderm gives support and some protection to the connecting sections and to the lower parts of individual polyps. Stony corals secrete massive calcium carbonate exoskeletons. A few polyps collect materials such as sand grains and shell fragments, which they attach to their outsides. Some colonial sea anemones stiffen the mesoglea with sediment particles.

Main Cell Layers

Cnidaria are diploblastic animals; in other words, they have two main cell layers, while more complex animals are triploblasts having three main layers. The two main cell layers of cnidarians form epithelia that are mostly one cell thick, and are attached to a fibrous basement membrane, which they secrete. They also secrete the jelly-like mesoglea that separates the layers. The layer that faces outwards, known as the ectoderm ("outside skin"), generally contains the following types of cells:

- Epitheliomuscular cells whose bodies form part of the epithelium but whose bases extend to form muscle fibers in parallel rows. The fibers of the outward-facing cell layer generally run at right angles to the fibers of the inward-facing one. In Anthozoa (anemones, corals, etc.) and Scyphozoa (jellyfish), the mesoglea also contains some muscle cells.

- Cnidocytes, the harpoon-like "nettle cells" that give the phylum Cnidaria its name. These appear between or sometimes on top of the muscle cells.

- Nerve cells. Sensory cells appear between or sometimes on top of the muscle cells, and communicate via synapses (gaps across which chemical signals flow) with motor nerve cells, which lie mostly between the bases of the muscle cells.

- Interstitial cells, which are unspecialized and can replace lost or damaged cells by transforming into the appropriate types. These are found between the bases of muscle cells.

In addition to epitheliomuscular, nerve and interstitial cells, the inward-facing gastroderm ("stomach skin") contains gland cells that secrete digestive enzymes. In some species it also contains low concentrations of cnidocytes, which are used to subdue prey that is still struggling.

The mesoglea contains small numbers of amoeba-like cells, and muscle cells in some species. However, the number of middle-layer cells and types are much lower than in sponges.

Polymorphism

Polymorphism refers to the occurrence of structurally and functionally more than two different types of individuals within the same organism. It is a characteristic feature of Cnidarians, particularly the polyp and medusa forms, or of zooids within colonial organisms like those in Hydrozoa. In Hydrozoans, colonial individuals arising from individualss zooids will take on separate tasks. For example, in *Obelia* there are feeding individuals, the gastrozooids; the individuals capable of asexual reproduction only, the gonozooids, blastostyles and free-living or sexually reproducing individuals, the medusae.

Cnidocytes

These "nettle cells" function as harpoons, since their payloads remain connected to the bodies of the cells by threads. Three types of cnidocytes are known:

- Nematocysts inject venom into prey, and usually have barbs to keep them embedded in the victims. Most species have nematocysts.

- Spirocysts do not penetrate the victim or inject venom, but entangle it by means of small sticky hairs on the thread.

- Ptychocysts are not used for prey capture — instead the threads of discharged ptychocysts are used for building protective tubes in which their owners live. Ptychocysts are found only in the order Ceriantharia, tube anemones.

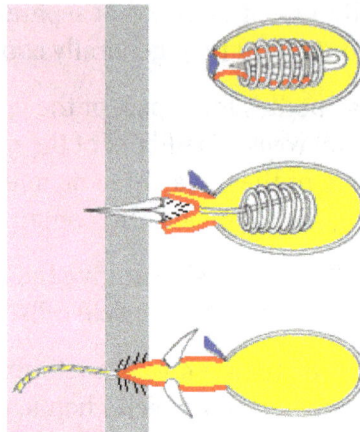

Firing sequence of the cnida in a hydra's nematocyst

■	Operculum (lid)	■	"Finger" that turns inside out
///	Barbs	■	Venom
■	Victim's skin	■	Victim's tissues

The main components of a cnidocyte are:

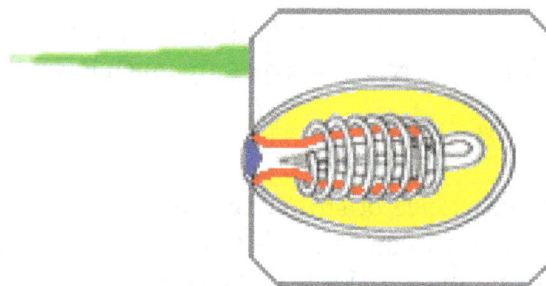

A hydra's nematocyst, before firing.
■ "trigger" cilium

- A cilium (fine hair) which projects above the surface and acts as a trigger. Spirocysts do not have cilia.

- A tough capsule, the cnida, which houses the thread, its payload and a mixture of chemicals that may include venom or adhesives or both. ("cnida" is derived from the Greek word κνίδη, which means "nettle")

- A tube-like extension of the wall of the cnida that points into the cnida, like the finger of a rubber glove pushed inwards. When a cnidocyte fires, the finger pops out. If the cell is a venomous nematocyte, the "finger"'s tip reveals a set of barbs that anchor it in the prey.

- The thread, which is an extension of the "finger" and coils round it until the cnidocyte fires. The thread is usually hollow and delivers chemicals from the cnida to the target.

- An operculum (lid) over the end of the cnida. The lid may be a single hinged flap or three flaps arranged like slices of pie.

- The cell body, which produces all the other parts.

- It is difficult to study the firing mechanisms of cnidocytes as these structures are small but very complex. At least four hypotheses have been proposed:

- Rapid contraction of fibers round the cnida may increase its internal pressure.

- The thread may be like a coiled spring that extends rapidly when released.

- In the case of *Chironex* (the "sea wasp"), chemical changes in the cnida's contents may cause them to expand rapidly by polymerization.

- Chemical changes in the liquid in the cnida make it a much more concentrated solution, so that osmotic pressure forces water in very rapidly to dilute it. This mechanism has been observed in nematocysts of the class Hydrozoa, sometimes producing pressures as high as 140 atmospheres, similar to that of scuba air tanks, and fully extending the thread in as little as 2 milliseconds (0.002 second).

Cnidocytes can only fire once, and about 25% of a hydra's nematocysts are lost from its tentacles when capturing a brine shrimp. Used cnidocytes have to be replaced, which takes about 48 hours. To minimise wasteful firing, two types of stimulus are generally required to trigger cnidocytes: nearby sensory cells detect chemicals in the water, and their cilia respond to contact. This combination prevents them from firing at distant or non-living objects. Groups of cnidocytes are usually connected by nerves and, if one fires, the rest of the group requires a weaker minimum stimulus than the cells that fire first.

Locomotion

Medusae swim by a form of jet propulsion: muscles, especially inside the rim of the bell, squeeze water out of the cavity inside the bell, and the springiness of the mesoglea powers the recovery stroke. Since the tissue layers are very thin, they provide too little power to swim against currents and just enough to control movement within currents.

Hydras and some sea anemones can move slowly over rocks and sea or stream beds by various means: creeping like snails, crawling like inchworms, or by somersaulting. A few can swim clumsily by waggling their bases.

Nervous System and Senses

Cnidarians are generally thought to have no brains or even central nervous systems. However, they do have integrative areas of neural tissue that could be considered some form of centralization. Most of their bodies are innervated by decentralized nerve nets that control their swimming musculature and connect with sensory structures, though clade has slightly different structures. These sensory structures, usually called rhopalia, can generate signals in response to various types of stimuli such as light, pressure, and much more. Medusa usually have several of them around the margin of the bell that work together to control the motor nerve net, that directly innervates the swimming muscles. Most Cnidarians also have a parallel system. In scyphozoans, this takes the form of a diffuse nerve net, which has modulatory effects on the nervous system. As well as forming the "signal cables" between sensory neurons and motoneurons, intermediate neurons in the nerve net can also form ganglia that act as local coordination centers. Communication between nerve cells can occur by chemical synapses or gap junctions in hydrozoans, though gap junctions are not present in all groups. Cnidarians have many of the same neurotransmitters as many animals, including chemicals such as glutamate, GABA, and acetylcholine.

This structure ensures that the musculature is excited rapidly and simultaneously, and can be directly stimulated from any point on the body, and it also is better able to recover after injury.

Medusae and complex swimming colonies such as siphonophores and chondrophores sense tilt and acceleration by means of statocysts, chambers lined with hairs which detect the movements of internal mineral grains called statoliths. If the body tilts in the wrong direction, the animal rights itself by increasing the strength of the swimming movements on the side that is too low. Most species have ocelli ("simple eyes"), which can detect sources of light. However the agile Box Jellyfish are unique among Medusae because they possess four kinds of true eyes that have retinas, corneas and lenses. Although the eyes probably do not form images, Cubozoa can clearly distinguish the direction from which light is coming as well as negotiate around solid-colored objects.

Feeding and Excretion

Cnidarians feed in several ways: predation, absorbing dissolved organic chemicals, filtering food particles out of the water, obtaining nutrients from symbiotic algae within their cells, and parasitism. Most obtain the majority of their food from predation but some, including the corals *Hetroxenia* and *Leptogorgia*, depend almost completely on their endosymbionts and on absorbing dissolved nutrients. Cnidaria give their symbiotic algae carbon dioxide, some nutrients, a place in the sun and protection against predators.

Predatory species use their cnidocytes to poison or entangle prey, and those with venomous nematocysts may start digestion by injecting digestive enzymes. The "smell" of fluids from wounded prey makes the tentacles fold inwards and wipe the prey off into the mouth. In medusae the tentacles round the edge of the bell are often short and most of the prey capture is done by "oral arms", which are extensions of the edge of the mouth and are often frilled and sometimes branched to increase their surface area. Medusae often trap prey or suspended food particles by swimming upwards, spreading their tentacles and oral arms and then sinking. In species for which suspended food particles are important, the tentacles and oral arms often have rows of cilia whose beating creates currents that flow towards the mouth, and some produce nets of mucus to trap particles. Their digestion is both intra and extracellular.

Once the food is in the digestive cavity, gland cells in the gastroderm release enzymes that reduce the prey to slurry, usually within a few hours. This circulates through the digestive cavity and, in colonial cnidarians, through the connecting tunnels, so that gastroderm cells can absorb the nutrients. Absorption may take a few hours, and digestion within the cells may take a few days. The circulation of nutrients is driven by water currents produced by cilia in the gastroderm or by muscular movements or both, so that nutrients reach all parts of the digestive cavity. Nutrients reach the outer cell layer by diffusion or, for animals or zooids such as medusae which have thick mesogleas, are transported by mobile cells in the mesoglea.

Indigestible remains of prey are expelled through the mouth. The main waste product of cells' internal processes is ammonia, which is removed by the external and internal water currents.

Respiration

There are no respiratory organs, and both cell layers absorb oxygen from and expel carbon dioxide into the surrounding water. When the water in the digestive cavity becomes stale it must be replaced, and nutrients that have not been absorbed will be expelled with it. Some Anthozoa have ciliated grooves on their tentacles, allowing them to pump water out of and into the digestive cavity without opening the mouth. This improves respiration after feeding and allows these animals, which use the cavity as a hydrostatic skeleton, to control the water pressure in the cavity without expelling undigested food.

Cnidaria that carry photosynthetic symbionts may have the opposite problem, an excess of oxygen, which may prove toxic. The animals produce large quantities of antioxidants to neutralize the excess oxygen.

Regeneration

All cnidarians can regenerate, allowing them to recover from injury and to reproduce asexually. Medusae have limited ability to regenerate, but polyps can do so from small pieces or even collections of separated cells. This enables corals to recover even after apparently being destroyed by predators.

Reproduction

Sexual

Cnidarian sexual reproduction often involves a complex life cycle with both polyp and medusa stages. For example, in Scyphozoa (jellyfish) and Cubozoa (box jellies) a larva swims until it finds a good site, and then becomes a polyp. This grows normally but then absorbs its tentacles and splits horizontally into a series of disks that become juvenile medusae, a process called strobilation. The juveniles swim off and slowly grow to maturity, while the polyp re-grows and may continue strobilating periodically. The adults have gonads in the gastroderm, and these release ova and sperm into the water in the breeding season.

This phenomenon of succession of differently organized generations (one asexually reproducing, sessile polyp, and one sexually reproducing, free-swimming medusa or sessile polyp) is sometimes called Alternation of asexual and sexual phases or metagenesis, but should not be confused with the alternation of generations as found in plants, characterized by an alternation of a multicellular spore-producing form and a multicellular gamete-producing form.

Shortened forms of this life cycle are common, for example some oceanic scyphozoans omit the polyp stage completely, and cubozoan polyps produce only one medusa. Hydrozoa have a variety of life cycles. Some have no polyp stages and some (e.g. *hydra*) have no medusae. In some species, the medusae remain attached to the polyp and are responsible for sexual reproduction; in extreme cases these reproductive zooids may not look much like medusae. Meanwhile, life cycle reversal, in which polyps are formed directly from medusae without the involvement of sexual reproduction process, was observed in both Hydrozoa (*Turritopsis dohrnii* and *Laodicea undulata*) and Scyphozoa (*Aurelia* sp.1). Anthozoa have no medusa stage at all and the polyps are responsible for sexual reproduction.

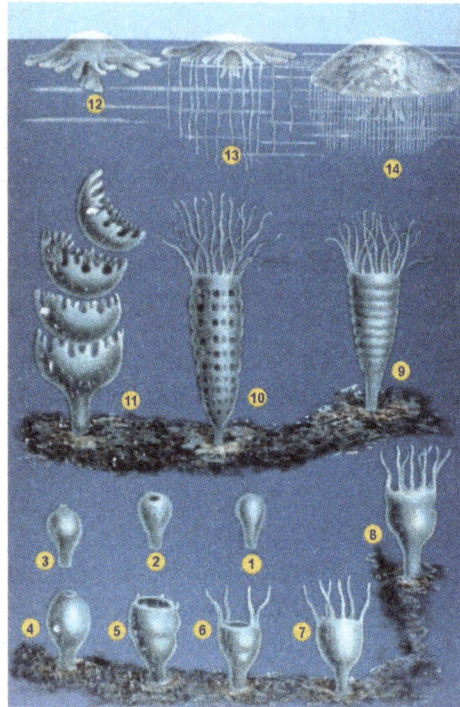

Life cycle of a jellyfish:
1–3 Larva searches for site
4–8 Polyp grows
9–11 Polyp strobilates
12–14 Medusa grows

Spawning is generally driven by environmental factors such as changes in the water temperature, and their release is triggered by lighting conditions such as sunrise, sunset or the phase of the moon. Many species of Cnidaria may spawn simultaneously in the same location, so that there are too many ova and sperm for predators to eat more than a tiny percentage — one famous example is the Great Barrier Reef, where at least 110 corals and a few non-cnidarian invertebrates produce enough gametes to turn the water cloudy. These mass spawnings may produce hybrids, some of which can settle and form polyps, but it is not known how long these can survive. In some species the ova release chemicals that attract sperm of the same species.

The fertilized eggs develop into larvae by dividing until there are enough cells to form a hollow sphere (blastula) and then a depression forms at one end (gastrulation) and eventually becomes the digestive cavity. However, in cnidarians the depression forms at the end further from the yolk (at the animal pole), while in bilaterians it forms at the other end (vegetal pole). The larvae, called planulae, swim or crawl by means of cilia. They are cigar-shaped but slightly broader at the "front" end, which is the aboral, vegetal-pole end and eventually attaches to a substrate if the species has a polyp stage.

Anthozoan larvae either have large yolks or are capable of feeding on plankton, and some already have endosymbiotic algae that help to feed them. Since the parents are immobile, these feeding capabilities extend the larvae's range and avoid overcrowding of sites. Scyphozoan and hydrozoan larvae have little yolk and most lack endosymbiotic algae, and therefore have to settle quickly and metamorphose into polyps. Instead, these species rely on their medusae to extend their ranges.

Asexual

All known cnidaria can reproduce asexually by various means, in addition to regenerating after being fragmented. Hydrozoan polyps only bud, while the medusae of some hydrozoans can divide down the middle. Scyphozoan polyps can both bud and split down the middle. In addition to both of these methods, Anthozoa can split horizontally just above the base. Asexual reproduction makes the daughter cnidarian a clone of the adult.

Classification

Cnidarians were for a long time grouped with Ctenophores in the phylum Coelenterata, but increasing awareness of their differences caused them to be placed in separate phyla. Modern cnidarians are generally classified into four main classes: sessile Anthozoa (sea anemones, corals, sea pens); swimming Scyphozoa (jellyfish) and Cubozoa (box jellies); and Hydrozoa, a diverse group that includes all the freshwater cnidarians as well as many marine forms, and has both sessile members such as *Hydra* and colonial swimmers such as the Portuguese Man o' War. Staurozoa have recently been recognised as a class in their own right rather than a sub-group of Scyphozoa, and the parasitic Myxozoa and Polypodiozoa are now recognized as highly derived cnidarians rather than more closely related to the bilaterians.

	Hydrozoa	Scyphozoa	Cubozoa	Anthozoa
Number of species	3,600	228	42	6,100
Examples	*Hydra*, siphono-phores	Jellyfish	Box jellies	Sea anemones, corals, sea pens
Cells found in mesoglea	No	Yes	Yes	Yes
Nematocysts in exodermis	No	Yes	Yes	Yes
Medusa phase in life cycle	In some species	Yes	Yes	No
Number of medusae produced per polyp	Many	Many	One	(not applicable)

Stauromedusae, small sessile cnidarians with stalks and no medusa stage, have traditionally been classified as members of the Scyphozoa, but recent research suggests they should be regarded as a separate class, Staurozoa.

The Myxozoa, microscopic parasites, were first classified as protozoans. Research then suggested that *Polypodium hydriforme*, a parasite *within* the egg cells of sturgeon, is closely related to the Myxozoa and that both *Polypodium* and the Myxozoa were intermediate between cnidarians and bilaterian animals. More recent research demonstrated that the previous identification of bilaterian genes reflected contamination of the Myxozoan samples by material from their host organism, and they are now firmly identified as heavily derived cnidarians, and more closely related to Hydrozoa and Scyphozoa than to Anthozoa.

Some researchers classify the extinct conulariids as cnidarians, while others propose that they form a completely separate phylum.

Ecology

Many cnidarians are limited to shallow waters because they depend on endosymbiotic algae for much of their nutrients. The life cycles of most have polyp stages, which are limited to locations that offer stable substrates. Nevertheless, major cnidarian groups contain species that have escaped these limitations. Hydrozoans have a worldwide range: some, such as *Hydra*, live in freshwater; *Obelia* appears in the coastal waters of all the oceans; and *Liriope* can form large shoals near the surface in mid-ocean. Among anthozoans, a few scleractinian corals, sea pens and sea fans live in deep, cold waters, and some sea anemones inhabit polar seabeds while others live near hydrothermal vents over 10 km (6.2 mi) below sea-level. Reef-building corals are limited to tropical seas between 30°N and 30°S with a maximum depth of 46 m (151 ft), temperatures between 20 °C (68 °F) and 28 °C (82 °F) high salinity and low carbon dioxide levels. Stauromedusae, although usually classified as jellyfish, are stalked, sessile animals that live in cool to Arctic waters. Cnidarians range in size from a mere handful of cells for myxozoan through *Hydra's* 5–20 mm (0.20–0.79 in) long, to the Lion's mane jellyfish, which may exceed 2 m (6.6 ft) in diameter and 75 m (246 ft) in length.

Prey of cnidarians ranges from plankton to animals several times larger than themselves. Some cnidarians are parasites, mainly on jellyfish but a few are major pests of fish. Others obtain most of their nourishment from endosymbiotic algae or dissolved nutrients. Predators of cnidarians include: sea slugs, which can incorporate nematocysts into their own bodies for self-defense; starfish, notably the crown of thorns starfish, which can devastate corals; butterfly fish and parrot fish, which eat corals; and marine turtles, which eat jellyfish. Some sea anemones and jellyfish have a symbiotic relationship with some fish; for example clown fish live among the tentacles of sea anemones, and each partner protects the other against predators.

Coral reefs form some of the world's most productive ecosystems. Common coral reef cnidarians include both Anthozoans (hard corals, octocorals, anemones) and Hydrozoans (fire corals, lace corals). The endosymbiotic algae of many cnidarian species are very effective primary producers, in other words converters of inorganic chemicals into organic ones that other organisms can use, and their coral hosts use these organic chemicals very efficiently. In addition, reefs provide complex and varied habitats that support a wide range of other organisms. Fringing reefs just below low-tide level also have a mutually beneficial relationship with mangrove forests at high-tide level and sea grass meadows in between: the reefs protect the mangroves and seagrass from strong currents and waves that would damage them or erode the sediments in which they are rooted, while the mangroves and seagrass protect the coral from large influxes of silt, fresh water and pollutants. This additional level of variety in the environment is beneficial to many types of coral reef animals, which for example may feed in the sea grass and use the reefs for protection or breeding.

Evolutionary History

Fossil Record

The earliest widely accepted animal fossils are rather modern-looking cnidarians, possibly from around 580 million years ago, although fossils from the Doushantuo Formation can only be dated approximately. The identification of some of these as embryos of animals has been contested, but other fossils from these rocks strongly resemble tubes and other mineralized structures made by corals. Their presence implies that the cnidarian and bilaterian lineages had already diverged.

Although the Ediacaran fossil *Charnia* used to be classified as a jellyfish or sea pen, more recent study of growth patterns in *Charnia* and modern cnidarians has cast doubt on this hypothesis, leaving only the Canadian polyp, *Haootia*, as the only bona-fide cnidarian body fossil from the Ediacaran. Few fossils of cnidarians without mineralized skeletons are known from more recent rocks, except in lagerstätten that preserved soft-bodied animals.

The fossil coral *Cladocora* from Pliocene rocks in Cyprus

A few mineralized fossils that resemble corals have been found in rocks from the Cambrian period, and corals diversified in the Early Ordovician. These corals, which were wiped out in the Permian-Triassic extinction about 251 million years ago, did not dominate reef construction since sponges and algae also played a major part. During the Mesozoic era rudist bivalves were the main reef-builders, but they were wiped out in the Cretaceous–Paleogene extinction event 65 million years ago, and since then the main reef-builders have been scleractinian corals.

Family Tree

It is difficult to reconstruct the early stages in the evolutionary "family tree" of animals using only morphology (their shapes and structures), because the large differences between Porifera (sponges), Cnidaria plus Ctenophora (comb jellies), Placozoa and Bilateria (all the more complex animals) make comparisons difficult. Hence reconstructions now rely largely or entirely on molecular phylogenetics, which groups organisms according to similarities and differences in their biochemistry, usually in their DNA or RNA.

It is now generally thought that the Calcarea (sponges with calcium carbonate spicules) are more closely related to Cnidaria, Ctenophora (comb jellies) and Bilateria (all the more complex animals) than they are to the other groups of sponges. In 1866 it was proposed that Cnidaria and Ctenophora were more closely related to each other than to Bilateria and formed a group called Coelenterata ("hollow guts"), because Cnidaria and Ctenophora both rely on the flow of water in and out of a single cavity for feeding, excretion and respiration. In 1881, it was proposed that Ctenophora and Bilateria were more closely related to each other, since they shared features that Cnidaria lack, for example muscles in the middle layer (mesoglea in Ctenophora, mesoderm in Bilateria). However more recent analyses indicate that these similarities are rather vague, and the current view, based on molecular phylogenetics, is that Cnidaria and Bilateria are more closely related to each other than either is to Ctenophora. This grouping of Cnidaria and Bilateria has been labelled "Planulozoa" because it suggests that the earliest Bilateria were similar to the planula larvae of Cnidaria.

Within the Cnidaria, the Anthozoa (sea anemones and corals) are regarded as the sister-group of the rest, which suggests that the earliest cnidarians were sessile polyps with no medusa stage. However, it is unclear how the other groups acquired the medusa stage, since Hydrozoa form medusae by budding from the side of the polyp while the other Medusozoa do so by splitting them off from the tip of the polyp. The traditional grouping of Scyphozoa included the Staurozoa, but morphology and molecular phylogenetics indicate that Staurozoa are more closely related to Cubozoa (box jellies) than to other "Scyphozoa". Similarities in the double body walls of Staurozoa and the extinct Conulariida suggest that they are closely related. The position of Anthozoa nearest the beginning of the cnidarian family tree also implies that Anthozoa are the cnidarians most closely related to Bilateria, and this is supported by the fact that Anthozoa and Bilateria share some genes that determine the main axes of the body.

However, in 2005 Katja Seipel and Volker Schmid suggested that cnidarians and ctenophores are simplified descendants of triploblastic animals, since ctenophores and the medusa stage of some cnidarians have striated muscle, which in bilaterians arises from the mesoderm. They did not commit themselves on whether bilaterians evolved from early cnidarians or from the hypothesized triploblastic ancestors of cnidarians.

In molecular phylogenetics analyses from 2005 onwards, important groups of developmental genes show the same variety in cnidarians as in chordates. In fact cnidarians, and especially anthozoans (sea anemones and corals), retain some genes that are present in bacteria, protists, plants and fungi but not in bilaterians.

The mitochondrial genome in the medusozoan cnidarians, unlike those in other animals, is linear with fragmented genes. The reason for this difference is unknown.

Interaction with Humans

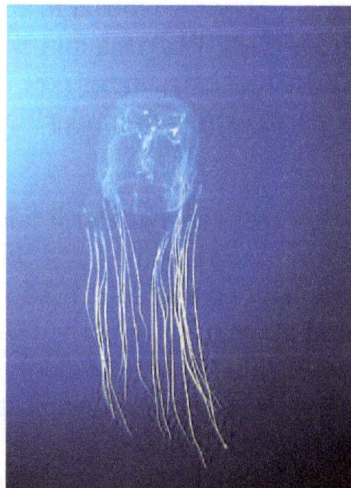

The dangerous "sea wasp" *Chironex fleckeri*

Jellyfish stings killed about 1,500 people in the 20th century, and cubozoans are particularly dangerous. On the other hand, some large jellyfish are considered a delicacy in East and Southeast Asia. Coral reefs have long been economically important as providers of fishing grounds, protectors of shore buildings against currents and tides, and more recently as centers of tourism. However, they are vulnerable to over-fishing, mining for construction materials, pollution, and damage caused by tourism.

Beaches protected from tides and storms by coral reefs are often the best places for housing in tropical countries. Reefs are an important food source for low-technology fishing, both on the reefs themselves and in the adjacent seas. However, despite their great productivity, reefs are vulnerable to over-fishing, because much of the organic carbon they produce is exhaled as carbon dioxide by organisms at the middle levels of the food chain and never reaches the larger species that are of interest to fishermen. Tourism centered on reefs provides much of the income of some tropical islands, attracting photographers, divers and sports fishermen. However, human activities damage reefs in several ways: mining for construction materials; pollution, including large influxes of fresh water from storm drains; commercial fishing, including the use of dynamite to stun fish and the capture of young fish for aquariums; and tourist damage caused by boat anchors and the cumulative effect of walking on the reefs. Coral, mainly from the Pacific Ocean has long been used in jewellery, and demand rose sharply in the 1980s.

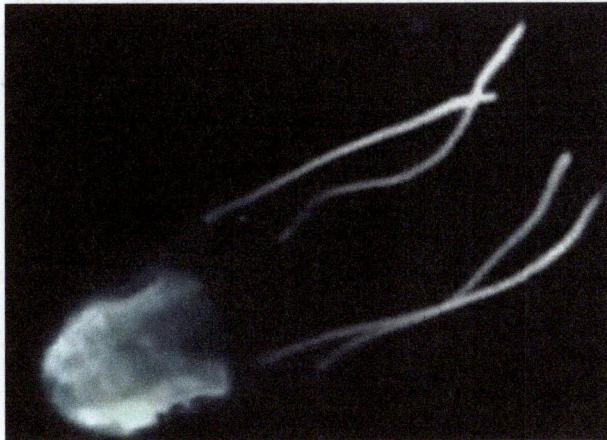

The dangerous *Carukia barnesi*, one of the known species of box jellyfish which can cause Irukandji syndrome.

Some large jellyfish species of the Rhizostomae order are commonly consumed in Japan, Korea and Southeast Asia. In parts of the range, fishing industry is restricted to daylight hours and calm conditions in two short seasons, from March to May and August to November. The commercial value of jellyfish food products depends on the skill with which they are prepared, and "Jellyfish Masters" guard their trade secrets carefully. Jellyfish is very low in cholesterol and sugars, but cheap preparation can introduce undesirable amounts of heavy metals.

The "sea wasp" *Chironex fleckeri* has been described as the world's most venomous jellyfish and is held responsible for 67 deaths, although it is difficult to identify the animal as it is almost transparent. Most stingings by *C. fleckeri* cause only mild symptoms. Seven other box jellies can cause a set of symptoms called Irukandji syndrome, which takes about 30 minutes to develop, and from a few hours to two weeks to disappear. Hospital treatment is usually required, and there have been a few deaths.

Tunicate

A tunicate is a marine invertebrate animal, a member of the subphylum Tunicata, which is part of the Chordata, a phylum which includes all animals with dorsal nerve cords and notochords. The subphylum was at one time called Urochordata, and the term urochordates is still sometimes used for these animals. Some tunicates live as solitary individuals, but others replicate by budding and become colonies, each unit being known as a zooid. They are marine filter feeders with a water-filled, sac-like body structure and two tubular openings, known as siphons, through which they draw in and expel water. During their respiration and feeding, they take in water through the incurrent (or inhalant) siphon and expel the filtered water through the excurrent (or exhalant) siphon. Most adult tunicates are sessile, and are permanently attached to rocks or other hard surfaces on the ocean floor; others, such as salps, doliolids and pyrosomes, swim in the pelagic zone of the sea as adults. Various species are commonly known as sea squirts, sea pork, sea livers, or sea tulips.

The earliest probable species of tunicate appears in the fossil record in the early Cambrian period. Despite their simple appearance and very different adult form, their close relationship to the vertebrates is evidenced by the fact that during their mobile larval stage, they possess a notochord or stiffening rod and resemble a tadpole. Their name derives from their unique outer covering or "tunic", which is formed from proteins and carbohydrates, and acts as an exoskeleton. In some species, it is thin, translucent, and gelatinous, while in others it is thick, tough, and stiff.

Taxonomy

Clavelina moluccensis, the bluebell tunicate

Botrylloides violaceus showing oral tentacles at openings of buccal siphons

About 2,150 species of tunicate exist in the world's oceans, living mostly in shallow water. The most numerous group is the ascidians; fewer than 100 species of these are found at depths greater than 200 m (660 ft). Some are solitary animals leading a sessile existence attached to the seabed, but others are colonial and a few are pelagic. Some are supported by a stalk, but most are attached directly to a substrate, which may be a rock, shell, coral, seaweed, mangrove root, dock, piling, or ship's hull. They are found in a range of solid or translucent colours and may resemble seeds, grapes, peaches, barrels, or bottles. One of the largest is a stalked sea tulip, *Pyura pachydermatina*, which can grow to be over 1 metre (3.3 ft) tall.

The Tunicata were established by Jean-Baptiste Lamarck in 1816. In 1881, Francis Maitland Balfour introduced a second name for the same group, "Urochorda", to emphasize the affinity of the group to other chordates. No doubt largely because of his influence, various authors supported the term, either as such, or as "Urochordata", but this usage is invalid because "Tunicata" has precedence, and grounds for superseding the name never existed. Accordingly, the current (formally correct) trend is to abandon the name Urochorda or Urochordata in favour of the original Tunicata, and the name Tunicata is almost invariably used in modern scientific works. It is accepted as valid by the World Register of Marine Species but not by the Integrated Taxonomic Information System.

Various common names are used for different species. Sea tulips are tunicates with colourful bodies supported on slender stalks. Sea squirts are so named because of their habit of contracting their bodies sharply and squirting out water when disturbed. Sea liver and sea pork get their names from the resemblance of their dead colonies to pieces of meat.

Classification

Tunicates are more closely related to craniates, (including hagfish, lampreys, and jawed vertebrates) than to lancelets, echinoderms, hemichordates, *Xenoturbella* or other invertebrates.

The clade consisting of tunicates and vertebrates is called Olfactores.

The Tunicata contain roughly 3,051 described species, traditionally divided into these classes:

- Ascidiacea (Aplousobranchia, Phlebobranchia, and Stolidobranchia)

- Thaliacea (Pyrosomida, Doliolida, and Salpida)

- Appendicularia (Larvacea)

Members of the Sorberacea were included in Ascidiacea in 2011 as a result of rDNA sequencing studies. Although the traditional classification is provisionally accepted, newer evidence suggests the Ascidiacea are an artificial group of paraphyletic status.

Fossil Record

Undisputed fossils of tunicates are rare. The best known and earliest unequivocally identified species is *Shankouclava shankouense* from the Lower Cambrian Maotianshan Shale at Shankou village, Anning, near Kunming (South China). There is also a common bioimmuration, (*Catellocaula vallata*), of a possible tunicate found in Upper Ordovician bryozoan skeletons of the upper midwestern United States.

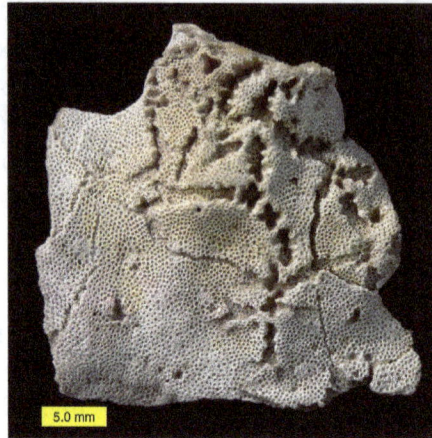

The star-shaped holes (*Catellocaula vallata*) in this Upper
Ordovician bryozoan may represent a tunicate preserved by bioimmuration in the bryozoan skeleton.

Three enigmatic species were also found from the Ediacaran period – *Ausia fenestrata* from the Nama Group of Namibia, the sac-like *Yarnemia acidiformis*, and one from a second new *Ausia*-like genus from the Onega Peninsula of northern Russia, *Burykhia hunti*. Results of a new study have shown possible affinity of these Ediacaran organisms to the ascidians. *Ausia* and *Burykhia* lived in shallow coastal waters slightly more than 555 to 548 million years ago, and are believed to be the oldest evidence of the chordate lineage of metazoans. The Russian Precambrian fossil *Yarnemia* is tentatively identified as a tunicate, although, because its fossils are nowhere near as well-preserved as those of *Ausia* and *Burykhia*, this identification is questioned.

Fossils of tunicates are rare because their bodies decay soon after death, but in some tunicate families, microscopic spicules are present, which may be preserved as microfossils. These spicules have occasionally been found in Jurassic and later rocks, but as few palaeontologists are familiar with them, they may have been mistaken for sponge spicules.

Hybridization Studies

A multi-taxon molecular study in 2010 proposed that sea squirts are descended from a hybrid between a chordate and a protostome ancestor. This study was based on a quartet partitioning approach designed to reveal horizontal gene transfer events among metazoan phyla

Anatomy

Body Form

Colonies of tunicates occur in a range of forms, and vary in the degree to which individual organisms, known as zooids, integrate with one another. In the simplest systems, the individual animals are widely separated, but linked together by horizontal connections called stolons, which grow along the seabed. Other species have the zooids growing closer together in a tuft or clustered together and sharing a common base. The most advanced colonies involve the integration of the zooids into a common structure surrounded by the tunic. These may have separate buccal siphons and a single central atrial siphon and may be organized into larger systems, with hundreds of star-shaped units. Often, the zooids in a colony are tiny but very numerous, and the colonies can form large encrusting or mat-like patches.

Colonial tunicate with multiple openings in a single tunic

Body Structure

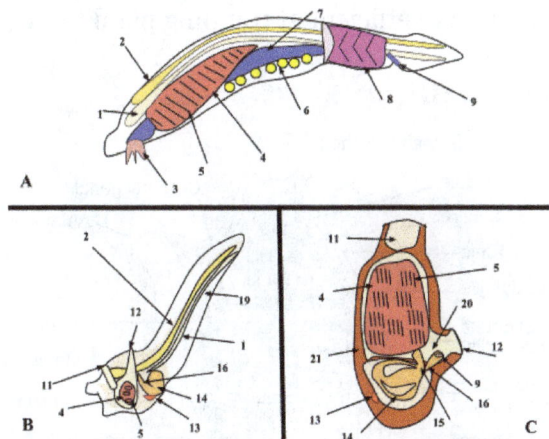

A. Lancelet, B. Larval tunicate, C. Adult tunicate. 1. Notochord, 2. Nerve chord, 3. Buccal cirri, 4. Pharynx, 5. Gill slit, 6. Gonad, 7. Gut, 8. V-shaped muscles, 9. Anus, 10. Inhalant syphon, 11. Exhalant syphon, 12. Heart, 13. Stomach, 14. Esophagus, 15. Intestines, 16. Tail, 17. Atrium, 18. Tunic.

By far the largest class of tunicates is the Ascidiacea. The body of an ascidiacean is surrounded by a test or tunic, from which the subphylum derives its name. This varies in thickness between species but may be tough, resembling cartilage, thin and delicate, or transparent and gelatinous. The tunic is composed of proteins and complex carbohydrates, and includes tunicin, a variety of cellulose. The tunic is unique among invertebrate exoskeletons in that it can grow as the animal enlarges and does not need to be periodically shed. Inside the tunic is the body wall or mantle composed of connective tissue, muscle fibres, blood vessels, and nerves. Two openings are found in the body wall: the buccal siphon at the top through which water flows into the interior, and the atrial siphon on the ventral side through which it is expelled. A large pharynx occupies most of the interior of the body. It is a muscular tube linking the buccal opening with the rest of the gut. It has a ciliated groove known as an endostyle on its ventral surface, and this secretes a mucous net which collects food particles and is wound up on the dorsal side of the pharynx. The gullet, at the lower end of the pharynx, links it to a loop of gut which terminates near the atrial siphon. The walls of the pharynx are perforated by several bands of slits, known as stigmata, through which water escapes into the surrounding water-filled cavity, the atrium. This is criss-crossed by various rope-like mesenteries which extend from the mantle and provide support for the pharynx, preventing it from collapsing, and also hold up the other organs.

The Thaliacea, the other main class of tunicates, is characterised by free-swimming, pelagic individuals. They are all filter feeders using a pharyngeal mucous net to catch their prey. The pyrosomes are bioluminous colonial tunicates with a hollow cylindrical structure. The buccal siphons are on the outside and the atrial siphons inside. About 10 species are known, and all are found in the tropics. The 23 species of doliolids are small, mostly under 2 cm (0.79 in) long. They are solitary, have the two siphons at opposite ends of their barrel-shaped bodies, and swim by jet propulsion. The 40 species of salps are also small, under 4 cm (1.6 in) long, and found in the surface waters of both warm and cold seas. They also move by jet propulsion, and often form long chains by budding off new individuals.

A third class, the Larvacea (or Appendicularia), is the only group of tunicates to retain their chordate characteristics in the adult state, a product of extensive neoteny. The 70 species of larvaceans superficially resemble the tadpole larvae of amphibians, although the tail is at right angles to the body. The notochord is retained, and the animals, mostly under 1 cm long, are propelled by undulations of the tail. They secrete an external mucous net known as a house, which may completely surround them and is very efficient at trapping planktonic particles.

Physiology and Internal Anatomy

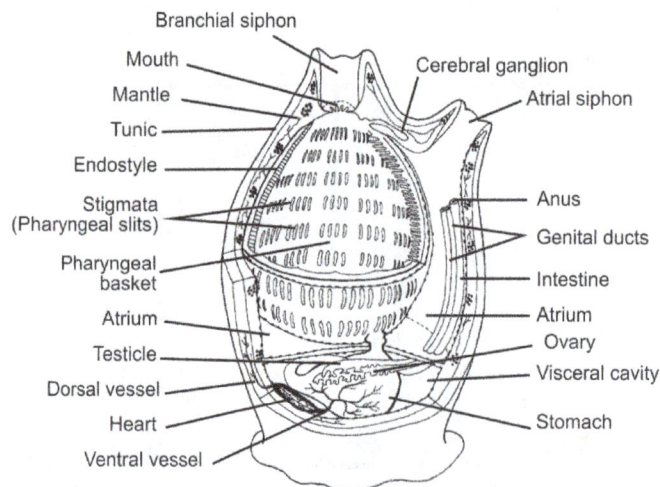

Internal anatomy of a generalised tunicate

Like other chordates, tunicates have a notochord during their early development, but by the time they have completed their larval stages, they have lost all myomeric segmentation throughout the body. As members of the Chordata, they are true Coelomata with endoderm, ectoderm, and mesoderm, but they do not develop very clear coelomic body cavities, if any at all. Whether they do or not, by the end of their larval development, all that remain are the pericardial, renal, and gonadal cavities of the adults. Except for the heart, gonads, and pharynx (or branchial sac), the organs are enclosed in a membrane called an epicardium, which is surrounded by the jelly-like mesenchyme. Tunicates begin life in a mobile larval stage that resembles a tadpole. A minority of species, those in the Larvacea, retain the general larval form throughout life, but most Tunicata very rapidly settle down and attach themselves to a suitable surface, later developing into a barrel-like and usually sedentary adult form. The Thaliacea, however, are pelagic throughout their lives and may have complex lifecycles.

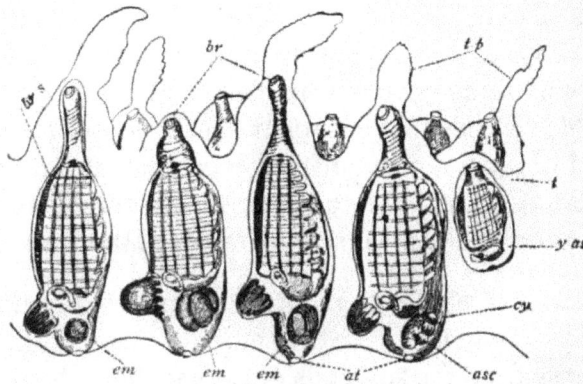

Section through the wall of a pyrosoma showing ascidiozooids; br) buccal siphon; at)
atrial siphon; tp) process of the test; br s) pharynx

Tunicates have a well-developed heart and circulatory system. The heart is a double U-shaped tube situated just below the gut. The blood vessels are simple connective tissue tubes, and their blood has several types of corpuscle. The blood may appear pale green, but this is not due to any respiratory pigments, and oxygen is transported dissolved in the plasma. Exact details of the circulatory system are unclear, but the gut, pharynx, gills, gonads, and nervous system seem to be arranged in series rather than in parallel, as happens in most other animals. Every few minutes, the heart stops beating and then restarts, pumping fluid in the reverse direction.

Tunicate blood has some unusual features. In some species of Ascidiidae and Perophoridae, it contains high concentrations of the transitional metal vanadium and vanadium-associated proteins in vacuoles in blood cells known as vanadocytes. Some tunicates can concentrate vanadium up to a level ten million times that of the surrounding seawater. It is stored in a +3 oxidation form that requires a pH of less than 2 for stability, and this is achieved by the vacuoles also containing sulphuric acid. The vanadocytes are later deposited just below the outer surface of the tunic, where their presence is thought to deter predation, although it is unclear whether this is due to the presence of the heavy metal or low pH. Other species of tunicates concentrate lithium, iron, niobium, and tantalum, which may serve a similar function. Other tunicate species produce distasteful organic compounds as chemical defenses against predators.

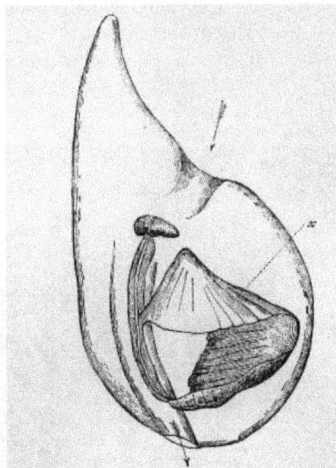

Oikopleura cophocerca in its "house". Arrows indicate water movement and (x) the lateral
reticulated parts of the house

Tunicates lack the kidney-like metanephridial organs typical of deuterostomes. Most have no excretory structures, but rely on the diffusion of ammonia across their tissues to rid themselves of nitrogenous waste, though some have a simple excretory system. The typical renal organ is a mass of large clear-walled vesicles that occupy the rectal loop, and the structure has no duct. Each vesicle is a remnant of a part of the primitive coelom, and its cells extract nitrogenous waste matter from circulating blood. They accumulate the wastes inside the vesicles as urate crystals, and do not have any obvious means of disposing of the material during their lifetimes.

Adult tunicates have a hollow cerebral ganglion, equivalent to a brain, and a hollow structure known as a neural gland. Both originate from the embryonic neural tube and are located between the two siphons. Nerves arise from the two ends of the ganglion; those from the anterior end innervate the buccal siphon and those from the posterior end supply the rest of the body, the atrial siphon, organs, gut and the musculature of the body wall. There are no sense organs but there are sensory cells on the siphons, the buccal tentacles and in the atrium.

Tunicates are unusual among animals in that they produce a large fraction of their tunic and some other structures in the form of cellulose. The production in animals of cellulose is so unusual that at first some researchers denied its presence outside of plants, but the tunicates were later found to possess a functional cellulose synthesizing enzyme, encoded by a gene horizontally transferred from a bacterium. When, in 1845, Carl Schmidt first announced the presence in the test of some ascidians of a substance very similar to cellulose, he called it "tunicine", but it is now recognized as cellulose rather than any alternative substance.

Blue sea squirts from the genus
Rhopalaea

Fluorescent-colored sea squirts,
Rhopalaea crassa

Didemnum molle

Feeding

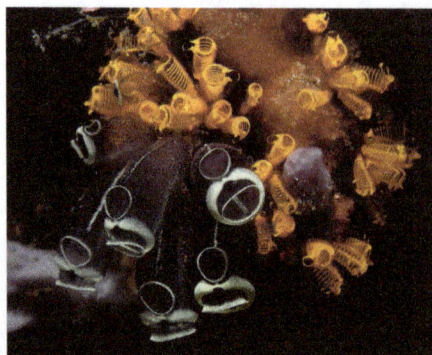

Clavelina robusta (black and white) and *Pycnoclavella flava* (orange) showing siphons

Nearly all tunicates are suspension feeders, capturing planktonic particles by filtering sea water through their bodies. Ascidians are typical in their digestive processes, but other tunicates have similar systems. Water is drawn into the body through the buccal siphon by the action of cilia lining the gill slits. To obtain enough food, an average ascidian needs to process one body-volume of water per second. This is drawn through a net lining the pharynx which is being continuously secreted by the endostyle. The net is made of sticky mucus threads with holes about 0.5 μm in diameter which can trap planktonic particles including bacteria. The net is rolled up on the dorsal side of the pharynx, and it and the trapped particles are drawn into the oesophagus. The gut is U-shaped and also ciliated to move the contents along. The stomach is an enlarged region at the lowest part of the U-bend. Here, digestive enzymes are secreted and a pyloric gland adds further secretions. After digestion, the food is moved on through the intestine, where absorption takes place, and the rectum, where undigested remains are formed into faecal pellets or strings. The anus opens into the dorsal or cloacal part of the peribranchial cavity near the atrial siphon. Here, the faeces are caught up by the constant stream of water which carries the waste to the exterior. The animal orientates itself to the current in such a way that the buccal siphon is always upstream and does not draw in contaminated water.

Some ascidians that live on soft sediments are detritivores. A few deepwater species, such as *Megalodicopia hians*, are sit-and-wait predators, trapping tiny crustacea, nematodes, and other small invertebrates with the muscular lobes which surround their buccal siphons. Certain tropical species in the family Didemnidae have symbiotic green algae or cyanobacteria in their tunics, and one of these symbionts, *Prochloron*, is unique to tunicates. Excess photosynthetic products are assumed to be available to the host.

Life Cycle

Ascidians are almost all hermaphrodites and each has a single ovary and testis, either near the gut or on the body wall. In some solitary species, sperm and eggs are shed into the sea and the larvae are planktonic. In others, especially colonial species, sperm is released into the water and drawn into the atria of other individuals with the incoming water current. Fertilization takes place here and the eggs are brooded through their early developmental stages. Some larval forms appear very much like primitive chordates with a notochord (stiffening rod) and superficially resemble small tadpoles. These swim by undulations of the tail and may have a simple eye, an ocellus, and a balancing organ, a statocyst.

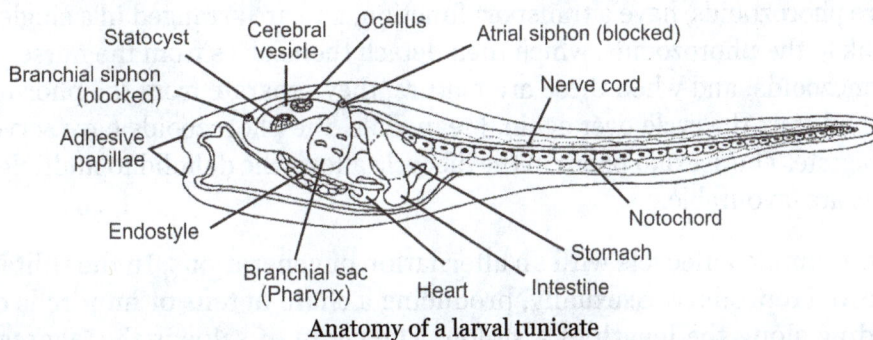

Anatomy of a larval tunicate

When sufficiently developed, the larva of the sessile species finds a suitable rock and cements itself in place. The larval form is not capable of feeding, though it may have a rudimentary digestive system, and is only a dispersal mechanism. Many physical changes occur to the tunicate's body during metamorphosis, one of the most significant being the reduction of the cerebral ganglion, which controls movement and is the equivalent of the vertebrate brain. From this comes the common saying that the sea squirt "eats its own brain". However, the adult does possess a cerebral ganglion which may even be larger than in the embrionic stage, so the scientific validity of this joke is questionable. In some classes, the adults remain pelagic (swimming or drifting in the open sea), although their larvae undergo similar metamorphoses to a higher or lower degree. Colonial forms also increase the size of the colony by budding off new individuals to share the same tunic.

Pyrosome colonies grow by budding off new zooids near the posterior end of the colony. Sexual reproduction starts within a zooid with an internally fertilized egg. This develops directly into an oozooid without any intervening larval form. This buds precociously to form four blastozooids which become detached in a single unit when the oozoid disintegrates. The atrial siphon of the oozoid becomes the exhalent siphon for the new, four-zooid colony.

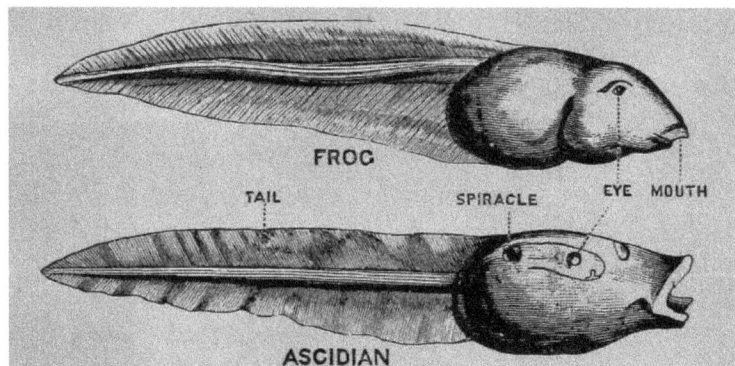
A 1901 comparison of frog tadpole and a tunicate larva

Doliolids have a very complex life cycle that includes various zooids with different functions. The sexually reproducing members of the colony are known as gonozooids. Each one is a hermaphrodite with the eggs being fertilised by sperm from another individual. The gonozooid is viviparous, and at first, the developing embryo feeds on its yolk sac before being released into the sea as a free-swimming, tadpole-like larva. This undergoes metamorphosis in the water column into an oozooid. This is known as a "nurse" as it develops a tail of zooids produced by budding asexually. Some of these are known as trophozooids, have a nutritional function, and are arranged in lateral rows. Others are phorozooids, have a transport function, and are arranged in a single central row. Other zooids link to the phorozooids, which then detach themselves from the nurse. These zooids develop into gonozooids, and when these are mature, they separate from the phorozooids to live independently and start the cycle over again. Meanwhile, the phorozooids have served their purpose and disintegrate. The asexual phase in the lifecycle allows the doliolid to multiply very rapidly when conditions are favourable.

Salps also have a complex lifecycle with an alternation of generations. In the solitary life history phase, an oozoid reproduces asexually, producing a chain of tens or hundreds of individual zooids by budding along the length of a stolon. The chain of salps is the 'aggregate' portion of the lifecycle. The aggregate individuals, known as blastozooids, remain attached together while swimming and feeding and growing larger. The blastozooids are sequential hermaphrodites. An egg in each is fertilized internally by a sperm from another colony. The egg develops in a brood sac inside the blastozooid and has a placental connection to the circulating blood of its "nurse". When it fills the blastozooid's body, it is released to start the independent life of an oozooid.

Larvaceans only reproduce sexually. They are protandrous hermaphrodites, except for *Oikopleura dioica* which is gonochoric, and a larva resembles the tadpole larva of ascidians. Once the trunk is fully developed, the larva undergoes "tail shift", in which the tail moves from a rearward position to a ventral orientation and twists through 90° relative to the trunk. The larva consists of a small, fixed number of cells, and grows by enlargement of these rather than cell division. Development is very rapid and only takes seven hours for a zygote to develop into a house-building juvenile starting to feed.

During embryonic development, tunicates exhibit determinate cleavage, where the fate of the cells is set early on with reduced cell numbers and genomes that are rapidly evolving. In contrast, the amphioxus and vertebrates show cell determination relatively late in development and cell cleavage is indeterminate. The genome evolution of amphioxus and vertebrates is also relatively slow.

Promotion of Out-crossing

Ciona intestinalis (class Ascidiacea) is a hermaphrodite that releases sperm and eggs into the surrounding seawater almost simultaneously. It is self-sterile, and thus has been used for studies on the mechanism of self-incompatibility. Self/non-self-recognition molecules play a key role in the process of interaction between sperm and the vitelline coat of the egg. It appears that self/non-self recognition in ascidians such as *C. intestinalis* is mechanistically similar to self-incompatibility systems in flowering plants. Self-incompatibility promotes out-crossing, and thus provides the adaptive advantage at each generation of the masking of deleterious recessive mutations (i.e. genetic complementation) and the avoidance of inbreeding depression.

Botryllus schlosseri (class Ascidiacea) is a colonial tunicate, a member of the only group of chordates that are able to reproduce both sexually and asexually. *B. schlosseri* is a sequential (protogynous) hermaphrodite, and in a colony, eggs are ovulated about two days before the peak of sperm emission. Thus self-fertilization is avoided, and cross-fertilization is favored. Although avoided, self-fertilization is still possible in *B. schlosseri*. Self-fertilized eggs develop with a substantially higher frequency of anomalies during cleavage than cross-fertilized eggs (23% vs. 1.6%). Also a significantly lower percentage of larvae derived from self-fertilized eggs metamorphose, and the growth of the colonies derived from their metamorphosis is significantly lower. These findings suggest that self-fertilization gives rise to inbreeding depression associated with developmental deficits that are likely caused by expression of deleterious recessive mutations.

A Model Tunicate

Oikopleura dioica (class Appendicularia) is a semelparous organism, reproducing only once in its lifetime. It employs an original reproductive strategy in which the entire female germ-line is contained within an ovary that is a single giant multinucleate cell termed the "coenocyst". *O. dioica* can be maintained in laboratory culture, and is of growing interest as a model organism because of its phylogenetic position within the closest sister group to vertebrates.

Invasive Species

Over the past few decades, tunicates (notably of the genera *Didemnum* and *Styela*) have been invading coastal waters in many countries. The carpet tunicate (*Didemnum vexillum*) has taken over a 6.5 sq mi (17 km²) area of the seabed on the Georges Bank off the northeast coast of North America, covering stones, molluscs, and other stationary objects in a dense mat. *D. vexillum*, *Styela clava* and *Ciona savignyi* have appeared and are thriving in Puget Sound and Hood Canal in the Pacific Northwest.

Invasive tunicates usually arrive as fouling organisms on the hulls of ships, but may also be introduced as larvae in ballast water. Another possible means of introduction is on the shells of molluscs brought in for marine cultivation. Current research indicates many tunicates previously thought to be indigenous to Europe and the Americas are, in fact, invaders. Some of these invasions may have occurred centuries or even millennia ago. In some areas, tunicates are proving to be a major threat to aquaculture operations.

Use by Humans

Medical Uses

Tunicates contain a host of potentially useful chemical compounds, including:

- Didemnins, effective against various types of cancer, as antivirals and as immunosuppressants

- Aplidine, a didemnin effective against various types of cancer

- Trabectedin, another didemnin effective against various types of cancer

Tunicates are able to correct their own cellular abnormalities over a series of generations, and a similar regenerative process may be possible for humans. The mechanisms underlying the phenomenon may lead to insights about the potential of cells and tissues to be reprogrammed and to regenerate compromised human organs.

As Food

Sea squirts for sale at a market, Busan, South Korea

Various Ascidiacea species are consumed as food around the world. In Japan and Korea, the sea pineapple (*Halocynthia roretzi*) is the main species eaten. It is cultivated on dangling cords made of palm fronds. In 1994, over 42,000 tons were produced, but since then, mass mortality events have occurred among the farmed sea squirts (the tunics becoming soft), and only 4,500 tons were produced in 2004.

Other Uses

The use of tunicates as a source of biofuel is being researched. The cellulose body wall can be broken down and converted into ethanol, and other parts of the animal are protein-rich and can be converted into fish feed. Culturing tunicates on a large scale may be possible and the economics of doing so are attractive. As tunicates have few predators, their removal from the sea may not have profound ecological impacts. Being sea-based, their production does not compete with food production as does the cultivation of land-based crops for biofuel projects.

Some tunicates are used as model organisms. *Ciona intestinalis* and *Ciona savignyi* have been used for developmental studies. Both species' mitochondrial and nuclear genomes have been sequenced. The nuclear genome of the appendicularian *Oikopleura dioica* appears to be one of the smallest among metazoans and this species has been used to study gene regulation and the evolution and development of chordates.

References

- Seilacher, Adolf (August 1992). "Vendobionta and Psammocorallia: lost constructions of Precambrian evolution". Journal of the Geological Society. London: Geological Society of London. 149 (4): 607–613. ISSN 0016-7649. doi:10.1144/gsjgs.149.4.0607. Retrieved 2015-02-04

- Ponder, W.F.; Lindberg, D.R., eds. (2008). Phylogeny and Evolution of the Mollusca. Berkeley: University of California Press. p. 481. ISBN 978-0-520-25092-5

- Davidson, Michael W. (May 26, 2005). "Animal Cell Structure". Molecular Expressions. Tallahassee, FL: Florida State University. Retrieved 2008-09-03

- Gewin, V (2005). "Functional genomics thickens the biological plot". PLOS Biology. 3 (6): e219. PMC 1149496. PMID 15941356. doi:10.1371/journal.pbio.0030219

- Jih-Pai Lin; Gon, Samuel M., III; Gehling, James G.; et al. (2006). "A Parvancorina-like arthropod from the Cambrian of South China". Historical Biology: An International Journal of Paleobiology. Taylor & Francis. 18 (1): 33–45. ISSN 1029-2381. doi:10.1080/08912960500508689

- Karleskint G, Richard Turner R and , James Small J (2012) Introduction to Marine Biology Cengage Learning, edition 4, page 445. ISBN 9781133364467

- "Where Does Our Head Come From? Brainless Sea Anemone Sheds New Light on the Evolutionary Origin of the Head". Science Daily. 2013-02-12. Retrieved 2014-01-18

- Lemaire, P (2011). "Evolutionary crossroads in developmental biology: the tunicates". Development. 138 (11): 2143–2152. PMID 21558365. doi:10.1242/dev.048975

- Robyn Crook & Jennifer Basil (2008). "A biphasic memory curve in the chambered nautilus, Nautilus pompilius L. (Cephalopoda: Nautiloidea)" (PDF). The Journal of Experimental Biology. 211 (12): 1992–1998. PMID 18515730. doi:10.1242/jeb.018531

- Platt HM (1994). "foreword". In Lorenzen S, Lorenzen SA. The phylogenetic systematics of freeliving nematodes. London: The Ray Society. ISBN 0-903874-22-9

- D. R. Currie; T. M. Ward (2009). South Australian Giant Crab (Pseudocarcinus gigas) Fishery (PDF). South

Australian Research and Development Institute. Fishery Assessment Report for PIRSA. Retrieved 9 December 2013

- Hart, M. W. (Jan–Feb 2002). "Life history evolution and comparative developmental biology of echinoderms". Evolutionary Development. 4 (1): 62–71. PMID 11868659. doi:10.1046/j.1525-142x.2002.01052

- Smith, J. E. (1937). "The structure and function of the tube feet in certain echinoderms" (PDF). Journal of the Marine Biological Association of the United Kingdom. 22 (1): 345–357. doi:10.1017/S0025315400012042

- Healy, J.M. (2001). "The Mollusca". In Anderson, D.T. Invertebrate Zoology (2 ed.). Oxford University Press. pp. 120–171. ISBN 0-19-551368-1

- Wray, Gregory A. (1999). "Echinodermata: Spiny-skinned animals: sea urchins, starfish, and their allies". Tree of Life web project. Retrieved 2012-10-19

- Sartori D., Gaion A. (2015). "Toxicity of polyunsaturated aldehydes of diatoms to Indo-Pacific bioindicator organism Echinometra mathaei". Drug and Chemical Toxicology: 1–5. doi:10.3109/01480545.2015.1041602

- David F. Wright (2017). "Bayesian estimation of fossil phylogenies and the evolution of early to middle Paleozoic crinoids (Echinodermata)". Journal of Paleontolgy. in press. doi:10.1017/jpa.2016.141

- Behrens, Peter; Bäuerlein, Edmund (2007). Handbook of Biomineralization: Biomimetic and bioinspired chemistry'. Wiley-VCH. p. 393. ISBN 3-527-31805-4

- Beamer, Victoria P. "Regeneration in Starfish (Asteroidea)". The Physiology of Arm Regeneration in Starfish (Asteroidea). Davidson College. Retrieved 2013-03-10

- Lieberman, BS (1999), "Testing the Darwinian Legacy of the Cambrian Radiation Using Trilobite Phylogeny and Biogeography", Journal of Paleontology, 73 (2): 176–181

- Raff, R A; Byrne, M (2006). "The active evolutionary lives of echinoderm larvae". Heredity. 97 (3): 244–52. PMID 16850040. doi:10.1038/sj.hdy.6800866

4

Phylum of Invertebrate Animal

Phylum is one of the categories in the hierarchy of biological classifications. Some of the types of phyla of invertebrates are flatworms, annelids, ctenophora and mollusca. The topics elaborated in this section will help in gaining a better perspective about the various phyla of invertebrates.

Ctenophora

Ctenophora (singular ctenophore, from the Greek 'comb' and 'carry'; commonly known as comb jellies) is a phylum of invertebrate animals that live in marine waters worldwide. Their most distinctive feature is the 'combs' – groups of cilia which they use for swimming – they are the largest animals that swim by means of cilia. Adults of various species range from a few millimeters to 1.5 m (4 ft 11 in) in size. Like cnidarians, their bodies consist of a mass of jelly, with one layer of cells on the outside and another lining the internal cavity. In ctenophores, these layers are two cells deep, while those in cnidarians are only one cell deep. Some authors combined ctenophores and cnidarians in one phylum, Coelenterata, as both groups rely on water flow through the body cavity for both digestion and respiration. Increasing awareness of the differences persuaded more recent authors to classify them as separate phyla.

Ctenophores also resemble cnidarians in having a decentralized nerve net rather than a brain. Genomic studies have suggested that the neurons of Ctenophora, which differ in many ways from other animal neurons, evolved independently from those of the other animals.

Almost all ctenophores are predators, taking prey ranging from microscopic larvae and rotifers to the adults of small crustaceans; the exceptions are juveniles of two species, which live as parasites on the salps on which adults of their species feed. In favorable circumstances, ctenophores can eat ten times their own weight in a day. Only 100–150 species have been validated, and possibly another 25 have not been fully described and named. The textbook examples are cydippids with egg-shaped bodies and a pair of retractable tentacles fringed with tentilla ("little tentacles") that are covered with colloblasts, sticky cells that capture prey. The phylum has a wide range of body forms, including the flattened, deep-sea platyctenids, in which the adults of most species lack combs, and the coastal beroids, which lack tentacles and prey on other ctenophores by using huge mouths armed with groups of large, stiffened cilia that act as teeth. These variations enable different species to build huge populations in the same area, because they specialize in different types of prey, which they capture by as wide a range of methods as spiders use.

Most species are hermaphrodites—a single animal can produce both eggs and sperm, meaning it can fertilize its own egg, not needing a mate. Some are simultaneous hermaphrodites, which can produce both eggs and sperm at the same time. Others are sequential hermaphrodites, in which the eggs and sperm mature at different times. Fertilization is generally external, although platyctenids' eggs are fertilized inside their parents' bodies and kept there until they hatch. The young are generally planktonic and in most species look like miniature cydippids, gradually changing into their adult shapes as they grow. The exceptions are the beroids, whose young are miniature beroids with large mouths and no tentacles, and the platyctenids, whose young live as cydippid-like plankton until they reach near-adult size, but then sink to the bottom and rapidly metamorphose into the adult form. In at least some species, juveniles are capable of reproduction before reaching the adult size and shape. The combination of hermaphroditism and early reproduction enables small populations to grow at an explosive rate.

Ctenophores may be abundant during the summer months in some coastal locations, but in other places they are uncommon and difficult to find. In bays where they occur in very high numbers, predation by ctenophores may control the populations of small zooplanktonic organisms such as copepods, which might otherwise wipe out the phytoplankton (planktonic plants), which are a vital part of marine food chains. One ctenophore, *Mnemiopsis*, has accidentally been introduced into the Black Sea, where it is blamed for causing fish stocks to collapse by eating both fish larvae and organisms that would otherwise have fed the fish. The situation was aggravated by other factors, such as over-fishing and long-term environmental changes that promoted the growth of the *Mnemiopsis* population. The later accidental introduction of *Beroe* helped to mitigate the problem, as *Beroe* preys on other ctenophores.

Despite their soft, gelatinous bodies, fossils thought to represent ctenophores, apparently with no tentacles but many more comb-rows than modern forms, have been found in lagerstätten as far back as the early Cambrian, about 515 million years ago. The position of the ctenophores in the evolutionary family tree of animals has long been debated, and the majority view at present, based on molecular phylogenetics, is that cnidarians and bilaterians are more closely related to each other than either is to ctenophores. A recent molecular phylogenetics analysis concluded that the common ancestor of all modern ctenophores was cydippid-like, and that all the modern groups appeared relatively recently, probably after the Cretaceous–Paleogene extinction event 66 million years ago. Evidence accumulating since the 1980s indicates that the "cydippids" are not monophyletic, in other words do not include all and only the descendants of a single common ancestor, because all the other traditional ctenophore groups are descendants of various cydippids.

Distinguishing Features

Ctenophores form an animal phylum that is more complex than sponges, about as complex as cnidarians (jellyfish, sea anemones, etc.), and less complex than bilaterians (which include almost all other animals). Unlike sponges, both ctenophores and cnidarians have: cells bound by inter-cell connections and carpet-like basement membranes; muscles; nervous systems; and some have sensory organs. Ctenophores are distinguished from all other animals by having colloblasts, which are sticky and adhere to prey, although a few ctenophore species lack them.

Like sponges and cnidarians, ctenophores have two main layers of cells that sandwich a middle layer of jelly-like material, which is called the mesoglea in cnidarians and ctenophores; more complex animals have three main cell layers and no intermediate jelly-like layer. Hence ctenophores and cnidarians have traditionally been labelled diploblastic, along with sponges. Both ctenophores and cnidarians have a type of muscle that, in more complex animals, arises from the middle cell layer, and as a result some recent text books classify ctenophores as triploblastic, while others still regard them as diploblastic.

Ranging from about 1 millimeter (0.039 in) to 1.5 meters (4.9 ft) in size, ctenophores are the largest non-colonial animals that use cilia ("hairs") as their main method of locomotion. Most species have eight strips, called comb rows, that run the length of their bodies and bear comb-like bands of cilia, called "ctenes", stacked along the comb rows so that when the cilia beat, those of each comb touch the comb below. The name "ctenophora" means "comb-bearing", from the Greek.

Comparison with other major animal groups				
	Sponges	**Cnidarians**	**Ctenophores**	**Bilateria**
Cnidocytes	No	Yes	Only in some species (obtained by ingesting cnidarians)	No
microRNA	Yes	Yes	No	Yes
Hox genes	No	No	Yes	Yes
Colloblasts	No		In most species	No
Digestive and circulatory organs	No			Yes
Number of main cell layers	Two, with jelly-like layer between them		Debate about whether two or three	Three
Cells in each layer bound together	No, except that Homoscleromorpha have basement membranes.	Yes: Inter-cell connections; basement membranes		
Sensory organs	No	Yes		
Number of cells in middle "jelly" layer	Many	Few		(Not applicable)
Cells in outer layers can move inwards and change functions	Yes	No		(Not applicable)
Nervous system	No	Yes, simple		Simple to complex
Muscles	None	Mostly epithelio-muscular	Mostly myoepithelial	Mostly myocytes

Description

For a phylum with relatively few species, ctenophores have a wide range of body plans. Coastal species need to be tough enough to withstand waves and swirling sediment particles, while some oceanic species are so fragile that it is very difficult to capture them intact for study. In addition oceanic species do not preserve well, and are known mainly from photographs and from observers' notes. Hence most attention has until recently concentrated on three coastal genera – *Pleurobrachia*, *Beroe* and *Mnemiopsis*. At least two textbooks base their descriptions of ctenophores on the cydippid *Pleurobrachia*.

Since the body of many species is *almost* radially symmetrical, the main axis is oral to aboral (from the mouth to the opposite end). However, since only two of the canals near the statocyst terminate in anal pores, ctenophores have no mirror-symmetry, although many have rotational symmetry, in other words if the animal rotates in a half-circle it looks the same as when it started.

Common Features

Body Layers

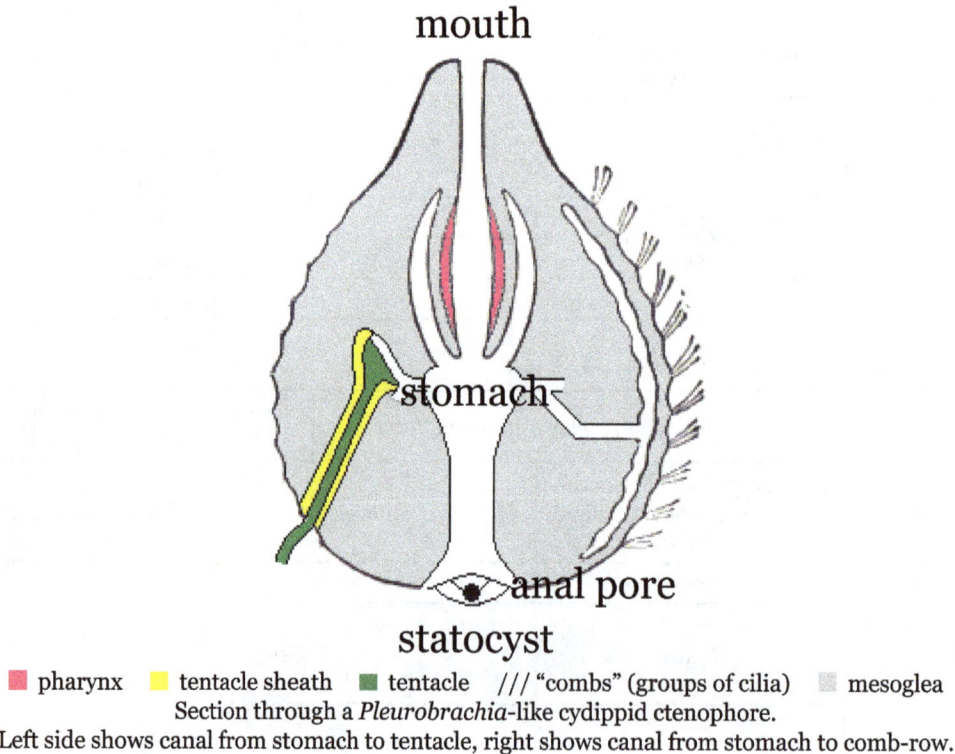

mouth

stomach

anal pore

statocyst

■ pharynx   ■ tentacle sheath   ■ tentacle   /// "combs" (groups of cilia)   ▨ mesoglea
Section through a *Pleurobrachia*-like cydippid ctenophore.
Left side shows canal from stomach to tentacle, right shows canal from stomach to comb-row.

Like those of cnidarians, (jellyfish, sea anemones, etc.), ctenophores' bodies consist of a relatively thick, jelly-like mesoglea sandwiched between two epithelia, layers of cells bound by inter-cell connections and by a fibrous basement membrane that they secrete. The epithelia of ctenophores have two layers of cells rather than one, and some of the cells in the upper layer have several cilia per cell.

The outer layer of the epidermis (outer skin) consists of: sensory cells; cells that secrete mucus, which protects the body; and interstitial cells, which can transform into other types of cell. In specialized parts of the body the outer layer also contains colloblasts, found along the surface of tentacles and used in capturing prey, or cells bearing multiple large cilia, for locomotion. The inner layer of the epidermis contains a nerve net, and myoepithelial cells that act as muscles.

The internal cavity forms: a mouth that can usually be closed by muscles; a pharynx ("throat"); a wider area in the center that acts as a stomach; and a system of internal canals. These branch through the mesoglea to the most active parts of the animal: the mouth and pharynx; the roots of the tentacles, if present; all along the underside of each comb row; and four branches round the sensory complex at the far end from the mouth – two of these four branches terminate in anal pores. The inner surface of the cavity is lined with an epithelium, the gastrodermis. The mouth and pharynx have both cilia and well-developed muscles. In other parts of the canal system, the gastrodermis is different on the sides nearest to and furthest from the organ that it supplies. The nearer side is composed of tall nutritive cells that store nutrients in vacuoles (internal compartments), germ cells that produce eggs or sperm, and photocytes that produce bioluminescence. The side furthest from the organ is covered with ciliated cells that circulate water through the canals, punctuated by ciliary rosettes, pores that are surrounded by double whorls of cilia and connect to the mesoglea.

Feeding, Excretion and Respiration

When prey is swallowed, it is liquefied in the pharynx by enzymes and by muscular contractions of the pharynx. The resulting slurry is wafted through the canal system by the beating of the cilia, and digested by the nutritive cells. The ciliary rosettes in the canals may help to transport nutrients to muscles in the mesoglea. The anal pores may eject unwanted small particles, but most unwanted matter is regurgitated via the mouth.

Little is known about how ctenophores get rid of waste products produced by the cells. The ciliary rosettes in the gastrodermis may help to remove wastes from the mesoglea, and may also help to adjust the animal's buoyancy by pumping water into or out of the mesoglea.

Locomotion

The outer surface bears usually eight comb rows, called swimming-plates, which are used for swimming. The rows are oriented to run from near the mouth (the "oral pole") to the opposite end (the "aboral pole"), and are spaced more or less evenly around the body, although spacing patterns vary by species and in most species the comb rows extend only part of the distance from the aboral pole towards the mouth. The "combs" (also called "ctenes" or "comb plates") run across each row, and each consists of thousands of unusually long cilia, up to 2 millimeters (0.079 in). Unlike conventional cilia and flagella, which has a filament structure arranged in a 9 + 2 pattern, these cilia are arranged in a 9 + 3 pattern, where the extra compact filament is suspected to have a supporting function. These normally beat so that the propulsion stroke is away from the mouth, although they can also reverse direction. Hence ctenophores usually swim in the direction in which the mouth is eating, unlike jellyfish. When trying to escape predators, one species can accelerate to six times its normal speed; some other species reverse direction as part of their escape behavior, by reversing the power stroke of the comb plate cilia.

It is uncertain how ctenophores control their buoyancy, but experiments have shown that some species rely on osmotic pressure to adapt to water of different densities. Their body fluids are normally as concentrated as seawater. If they enter less dense brackish water, the ciliary rosettes in the body cavity may pump this into the mesoglea to increase its bulk and decrease its density, to avoid sinking. Conversely if they move from brackish to full-strength seawater, the rosettes may pump water out of the mesoglea to reduce its volume and increase its density.

Nervous System and Senses

Ctenophores have no brain or central nervous system, but instead have a nerve net (rather like a cobweb) that forms a ring round the mouth and is densest near structures such as the comb rows, pharynx, tentacles (if present) and the sensory complex furthest from the mouth.

The largest single sensory feature is the aboral organ (at the opposite end from the mouth). Its main component is a statocyst, a balance sensor consisting of a statolith, a tiny grain of calcium carbonate, supported on four bundles of cilia, called "balancers", that sense its orientation. The statocyst is protected by a transparent dome made of long, immobile cilia. A ctenophore does not automatically try to keep the statolith resting equally on all the balancers. Instead its response is determined by the animal's "mood", in other words the overall state of the nervous system. For example, if a ctenophore with trailing tentacles captures prey, it will often put some comb rows into reverse, spinning the mouth towards the prey.

Cydippids

Aulacoctena sp., a cydippid ctenophore

Cydippid ctenophores have bodies that are more or less rounded, sometimes nearly spherical and other times more cylindrical or egg-shaped; the common coastal "sea gooseberry", *Pleurobrachia*, sometimes has an egg-shaped body with the mouth at the narrow end, although some individuals are more uniformly round. From opposite sides of the body extends a pair of long, slender tentacles, each housed in a sheath into which it can be withdrawn. Some species of cydippids have bodies that are flattened to various extents, so that they are wider in the plane of the tentacles.

The tentacles of cydippid ctenophores are typically fringed with tentilla ("little tentacles"), although a few genera have simple tentacles without these sidebranches. The tentacles and tentilla are densely covered with microscopic colloblasts that capture prey by sticking to it. Colloblasts are specialized mushroom-shaped cells in the outer layer of the epidermis, and have three main components: a domed head with vesicles (chambers) that contain adhesive; a stalk that anchors the cell in the lower layer of the epidermis or in the mesoglea; and a spiral thread that coils round the stalk and is attached to the head and to the root of the stalk. The function of the spiral thread is uncertain, but it may absorb stress when prey tries to escape, and thus prevent the collobast from being torn apart. In addition to colloblasts, members of the genus *Haeckelia*, which feed mainly on jellyfish, incorporate their victims' stinging nematocytes into their own tentacles – some cnidaria-eating nudibranchs similarly incorporate nematocytes into their bodies for defense. The tentilla of *Euplokamis* differ significantly from those of other cydippids: they contain striated muscle, a cell type otherwise unknown in the phylum Ctenophora; and they are coiled when relaxed, while the tentilla of all other known ctenophores elongate when relaxed. *Euplokamis'* tentilla have three types of movement that are used in capturing prey: they may flick out very quickly (in 40 to 60 milliseconds); they can wriggle, which may lure prey by behaving like small planktonic worms; and they coil round prey. The unique flicking is an uncoiling movement powered by contraction of the striated muscle. The wriggling motion is produced by smooth muscles, but of a highly specialized type. Coiling around prey is accomplished largely by the return of the tentilla to their inactive state, but the coils may be tightened by smooth muscle.

There are eight rows of combs that run from near the mouth to the opposite end, and are spaced evenly round the body. The "combs" beat in a metachronal rhythm rather like that of a Mexican wave. From each balancer in the statocyst a ciliary groove runs out under the dome and then splits to connect with two adjacent comb rows, and in some species runs all the way along the comb rows. This forms a *mechanical* system for transmitting the beat rhythm from the combs to the balancers, via water disturbances created by the cilia.

Lobates

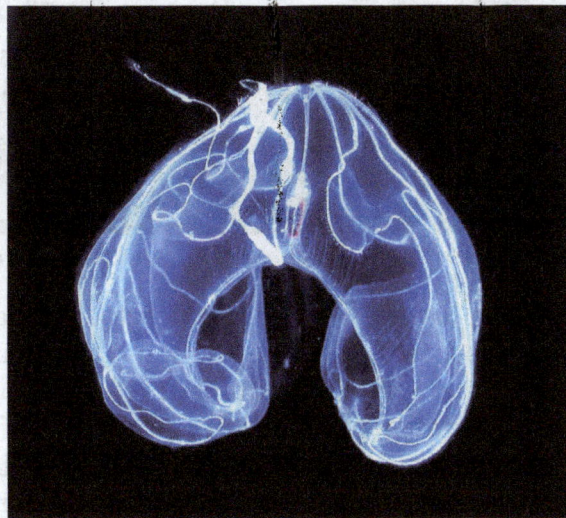

Bathocyroe fosteri a common but fragile deep-sea lobate, oriented mouth down.

The Lobata have a pair of lobes, which are muscular, cuplike extensions of the body that project beyond the mouth. Their inconspicuous tentacles originate from the corners of the mouth, running in convoluted grooves and spreading out over the inner surface of the lobes (rather than trailing far behind, as in the Cydippida). Between the lobes on either side of the mouth, many species of lobates have four auricles, gelatinous projections edged with cilia that produce water currents that help direct microscopic prey toward the mouth. This combination of structures enables lobates to feed continuously on suspended planktonic prey.

Lobates have eight comb-rows, originating at the aboral pole and usually not extending beyond the body to the lobes; in species with (four) auricles, the cilia edging the auricles are extensions of cilia in four of the comb rows. Most lobates are quite passive when moving through the water, using the cilia on their comb rows for propulsion, although *Leucothea* has long and active auricles whose movements also contribute to propulsion. Members of the lobate genera *Bathocyroe* and

Ocyropsis can escape from danger by clapping their lobes, so that the jet of expelled water drives them backwards very quickly. Unlike cydippids, the movements of lobates' combs are coordinated by nerves rather than by water disturbances created by the cilia, yet combs on the same row beat in the same Mexican wave style as the mechanically coordinated comb rows of cydippids and beroids. This may have enabled lobates to grow larger than cydippids and to have shapes that are less egg-like.

An unusual species first described in 2000, *Lobatolampea tetragona*, has been classified as a lobate, although the lobes are "primitive" and the body is medusa-like when floating and disk-like when resting on the sea-bed.

Beroids

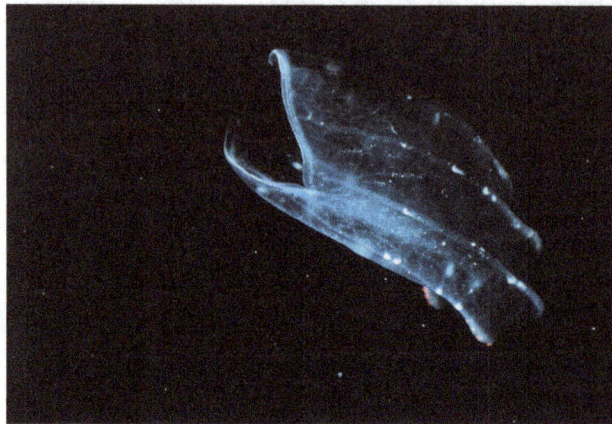

Beroe sp. swimming with open mouth, at left. This animal is 3–6 cm long.

The Beroida, also known as Nuda, have no feeding appendages, but their large pharynx, just inside the large mouth and filling most of the saclike body, bears "macrocilia" at the oral end. These fused bundles of several thousand large cilia are able to "bite" off pieces of prey that are too large to swallow whole – almost always other ctenophores. In front of the field of macrocilia, on the mouth "lips" in some species of *Beroe*, is a pair of narrow strips of adhesive epithelial cells on the stomach wall that "zip" the mouth shut when the animal is not feeding, by forming intercellular connections with the opposite adhesive strip. This tight closure streamlines the front of the animal when it is pursuing prey.

Other Body Forms

The Ganeshida have a pair of small oral lobes and a pair of tentacles. The body is circular rather than oval in cross-section, and the pharynx extends over the inner surfaces of the lobes.

The Thalassocalycida, only discovered in 1978 and known from only one species, are medusa-like, with bodies that are shortened in the oral-aboral direction, and short comb-rows on the surface furthest from the mouth, originating from near the aboral pole. They capture prey by movements of the bell and possibly by using two short tentacles.

The Cestida ("belt animals") are ribbon-shaped planktonic animals, with the mouth and aboral organ aligned in the middle of opposite edges of the ribbon. There is a pair of comb-rows along each aboral edge, and tentilla emerging from a groove all along the oral edge, which stream back across most of the wing-like body surface. Cestids can swim by undulating their bodies as well as by the beating of their comb-rows. There are two known species, with worldwide distribution in warm, and warm-temperate waters: *Cestum veneris* ("Venus' girdle") is among the largest ctenophores – up to 1.5 meters (4.9 ft) long, and can undulate slowly or quite rapidly. *Velamen parallelum*, which is typically less than 20 centimeters (0.66 ft) long, can move much faster in what has been described as a "darting motion".

Most Platyctenida have oval bodies that are flattened in the oral-aboral direction, with a pair of tentilla-bearing tentacles on the aboral surface. They cling to and creep on surfaces by everting the pharynx and using it as a muscular "foot". All but one of the known platyctenid species lack comb-rows. Platyctenids are usually cryptically colored, live on rocks, algae, or the body surfaces of other invertebrates, and are often revealed by their long tentacles with many sidebranches, seen streaming off the back of the ctenophore into the current.

Reproduction and Development

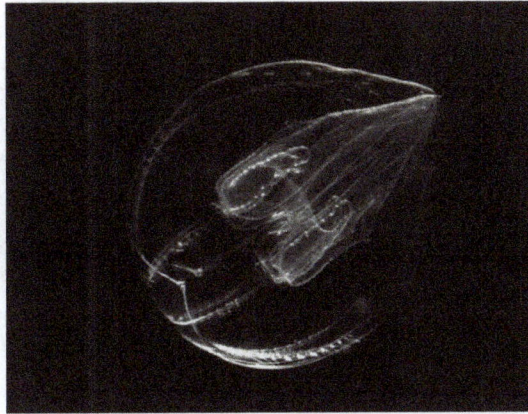

Cydippid larva of *Bolinopsis* sp., a few mm long.

Adults of most species can regenerate tissues that are damaged or removed, although only platyctenids reproduce by cloning, splitting off from the edges of their flat bodies fragments that develop into new individuals.

Almost all species are hermaphrodites, in other words they function as both males and females at the same time – except that in two species of the genus *Ocryopsis* individuals remain of the same single sex all their lives. The gonads are located in the parts of the internal canal network under the comb rows, and eggs and sperm are released via pores in the epidermis. Fertilization is external in most species, but platyctenids use internal fertilization and keep the eggs in brood chambers until they hatch. Self-fertilization has occasionally been seen in species of the genus *Mnemiopsis*, and it is thought that most of the hermaphroditic species are self-fertile.

Development of the fertilized eggs is direct, in other words there is no distinctive larval form, and juveniles of all groups generally resemble miniature cydippid adults. In the genus *Beroe* the juveniles, like the adults, lack tentacles and tentacle sheaths. In most species the juveniles gradually develop the body forms of their parents. In some groups, such as the flat, bottom-dwelling platyctenids, the juveniles behave more like true larvae, as they live among the plankton and thus occupy a different ecological niche from their parents and attain the adult form by a more radical metamorphosis, after dropping to the sea-floor.

At least in some species, juvenile ctenophores appear capable of producing small quantities of eggs and sperm while they are well below adult size, and adults produce eggs and sperm for as long as they have sufficient food. If they run short of food, they first stop producing eggs and sperm, and then shrink in size. When the food supply improves, they grow back to normal size and then resume reproduction. These features make ctenophores capable of increasing their populations very quickly.

Colors and Bioluminescence

Light diffracting along the comb rows of a *Mertensia ovum*. The left tentacle is deployed, the right one retracted.

Undescribed deep-sea species known as "Tortugas red", with trailing tentacles and clearly visible sidebranches, or tentilla.

Most ctenophores that live near the surface are mostly colorless and almost transparent. However some deeper-living species are strongly pigmented, for example the species known as "Tortugas red", which has not yet been formally described. Platyctenids generally live attached to other sea-bottom organisms, and often have similar colors to these host organisms. The gut of the deep-sea genus *Bathocyroe* is red, which hides the bioluminescence of copepods it has swallowed.

The comb rows of most planktonic ctenophores produce a rainbow effect, which is not caused by bioluminescence but by the scattering of light as the combs move. Most species are also bioluminescent, but the light is usually blue or green and can only be seen in darkness. However some significant groups, including all known platyctenids and the cydippid genus *Pleurobrachia*, are incapable of bioluminescence.

When some species, including *Bathyctena chuni*, *Euplokamis stationis* and *Eurhamphaea vexilligera*, are disturbed, they produce secretions (ink) that luminesce at much the same wavelengths as their bodies. Juveniles will luminesce more brightly in relation to their body size than adults, whose luminescence is diffused over their bodies. Detailed statistical investigation has not suggested the function of ctenophores' bioluminescence nor produced any correlation between its exact color and any aspect of the animals' environments, such as depth or whether they live in coastal or mid-ocean waters.

In ctenophores, bioluminescence is caused by the activation of calcium-activated proteins named photoproteins in cells called photocytes, which are often confined to the meridional canals that underlie the eight comb rows. In the genome of *Mnemiopsis leidyi* ten genes encode photoproteins. These genes are co-expressed with opsin genes in the developing photocytes of *Mnemiopsis leidyi*, raising the possibility that light production and light detection may be working together in these animals.

Ecology

Distribution

Ctenophores are found in most marine environments: from polar waters to the tropics; near coasts and in mid-ocean; from the surface waters to the ocean depths. The best-understood are the genera *Pleurobrachia*, *Beroe* and *Mnemiopsis*, as these planktonic coastal forms are among the most likely to be collected near shore. No ctenophores have been found in fresh water. In Cartagena a marine biologist , Fredy Orjuela, found ctenophores in brackish water between 20-28 upp. the uniqueness of his recent findings is that a ctenophore was found in LAGUITO, a closed and isolated water lagoon that used to be a small inner sea inning.

Prey and Predators

Almost all ctenophores are predators – there are no vegetarians and only one genus that is partly parasitic. If food is plentiful, they can eat 10 times their own weight per day. While *Beroe* preys mainly on other ctenophores, other surface-water species prey on zooplankton (planktonic animals) ranging in size from the microscopic, including mollusc and fish larvae, to small adult crustaceans such as copepods, amphipods, and even krill. Members of the genus *Haeckelia* prey on jellyfish and incorporate their prey's nematocysts (stinging cells) into their own tentacles instead of colloblasts. Ctenophores have been compared to spiders in their wide range of techniques from capturing prey – some hang motionless in the water using their tentacles as "webs", some are ambush predators like Salticid jumping spiders, and some dangle a sticky droplet at the end of a fine thread, as bolas spiders do. This variety explains the wide range of body forms in a phylum with rather few species. The two-tentacled "cydippid" *Lampea* feeds exclusively on salps, close relatives of sea-squirts that form large chain-like floating colonies, and juveniles of *Lampea* attach themselves like parasites to salps that are too large for them to swallow. Members of the cydippid genus *Pleurobrachia* and the lobate *Bolinopsis* often reach high population densities at the same place and time because they specialize in different types of prey: *Pleurobrachia*'s long tentacles mainly capture relatively strong swimmers such as adult copepods, while *Bolinopsis* generally feeds on smaller, weaker swimmers such as rotifers and mollusc and crustacean larvae.

Ctenophores used to be regarded as "dead ends" in marine food chains because it was thought their low ratio of organic matter to salt and water made them a poor diet for other animals. It is also often difficult to identify the remains of ctenophores in the guts of possible predators, although the combs sometimes remain intact long enough to provide a clue. Detailed investigation of chum salmon, *Oncorhynchus keta*, showed that these fish digest ctenophores 20 times as fast as an equal weight of shrimps, and that ctenophores can provide a good diet if there are enough of them around. Beroids prey mainly on other ctenophores. Some jellyfish and turtles eat large quantities of ctenophores, and jellyfish may temporarily wipe out ctenophore populations. Since ctenophores and jellyfish often have large seasonal variations in population, most fish that prey on them are generalists, and may have a greater effect on populations than the specialist jelly-eaters. This is underlined by an observation of herbivorous fishes deliberately feeding on gelatinous zooplankton during blooms in the Red Sea. The larvae of some sea anemones are parasites on ctenophores, as are the larvae of some flatworms that parasitize fish when they reach adulthood.

Ecological Impacts

Beroe ovata at the surface on the Black Sea coast

Ctenophores may balance marine ecosystems by preventing an over-abundance of copepods from eating all the phytoplankton (planktonic plants), which are the dominant marine producers of organic matter from non-organic ingredients.

On the other hand, in the late 1980s the Western Atlantic ctenophore *Mnemiopsis leidyi* was accidentally introduced into the Black Sea and Sea of Azov via the ballast tanks of ships, and has been blamed for causing sharp drops in fish catches by eating both fish larvae and small crustaceans that would otherwise feed the adult fish. *Mnemiopsis* is well equipped to invade new territories (although this was not predicted until after it so successfully colonized the Black Sea), as it can breed very rapidly and tolerate a wide range of water temperatures and salinities. The impact was increased by chronic overfishing, and by eutrophication that gave the entire ecosystem a short-term boost, causing the *Mnemiopsis* population to increase even faster than normal – and above all by the absence of efficient predators on these introduced ctenophores. *Mnemiopsis* populations in those areas were eventually brought under control by the accidental introduction of the *Mnemiopsis*-eating North American ctenophore *Beroe ovata*, and by a cooling of the local climate from 1991 to 1993, which significantly slowed the animal's metabolism. However the abundance of plankton in the area seems unlikely to be restored to pre-*Mnemiopsis* levels.

In the late 1990s *Mnemiopsis* appeared in the Caspian Sea. *Beroe ovata* arrived shortly after, and is expected to reduce but not eliminate the impact of *Mnemiopsis* there. *Mnemiopsis* also reached the eastern Mediterranean in the late 1990s and now appears to be thriving in the North Sea and Baltic Sea.

Classification

The number of known living ctenophore species is uncertain, since many of those named and formally described have turned out to be identical to species known under other scientific names.

Claudia Mills estimates that there about 100 to 150 valid species that are not duplicates, and that at least another 25, mostly deep-sea forms, have been recognized as distinct but not yet analyzed in enough detail to support a formal description and naming.

The traditional classification divides ctenophores into two classes, those with tentacles (Tentaculata) and those without (Nuda). The Nuda contains only one order (Beroida) and family (Beroidae), and two genera, *Beroe* (several species) and *Neis* (one species).

The Tentaculata are divided into the following eight orders:

- Cydippida, egg-shaped animals with long tentacles

- Lobata, with paired thick lobes

- Platyctenida, flattened animals that live on or near the sea-bed; most lack combs as adults, and use their pharynges as suckers to attach themselves to surfaces

- Ganeshida, with a pair of small lobes round the mouth, but an extended pharynx like that of platyctenids

- Cambojiida

- Cryptolobiferida

- Thalassocalycida, with short tentacles and a jellyfish-like "umbrella"

- Cestida, ribbon-shaped and the largest ctenophores

Evolutionary History

Fossil Record

Because of their soft, gelatinous bodies, ctenophores are extremely rare as fossils, and fossils that have been interpreted as ctenophores have been found only in lagerstätten, places where the environment was exceptionally suited to preservation of soft tissue. Until the mid-1990s only two specimens good enough for analysis were known, both members of the crown group, from the early Devonian (Emsian) period. Three additional putative species were then found in the Burgess Shale and other Canadian rocks of similar age, about 505 million years ago in the mid-Cambrian period. All three apparently lacked tentacles but had between 24 and 80 comb rows, far more than the 8 typical of living species. They also appear to have had internal organ-like structures unlike anything found in living ctenophores. One of the fossil species first reported in 1996 had a large mouth, apparently surrounded by a folded edge that may have been muscular. Evidence from China a year later suggests that such ctenophores were widespread in the Cambrian, but perhaps very different from modern species – for example one fossil's comb-rows were mounted on prominent vanes. The Ediacaran *Eoandromeda* could putatively represent a comb jelly.

The early Cambrian sessile frond-like fossil *Stromatoveris*, from China's Chengjiang lagerstätte and dated to about 515 million years ago, is very similar to Vendobionta of the preceding Ediacaran period. De-Gan Shu, Simon Conway Morris *et al.* found on its branches what they considered rows of cilia, used for filter feeding. They suggested that *Stromatoveris* was an evolutionary "aunt" of ctenophores, and that ctenophores originated from sessile animals whose descendants became swimmers and changed the cilia from a feeding mechanism to a propulsion system.

520 million years old Cambrian fossils also from Chengjiang in China show a now wholly extinct class of ctenophore, named "Scleroctenophora", that had a complex internal skeleton with long spines.

Relationship to other Animal Groups

The relationship of ctenophores to the rest of Metazoa is very important to our understanding of the early evolution of animals and the origin of multicellularity. It has been the focus of debate for many years. Ctenophores have been purported to be the sister lineage to the Bilateria, sister to the Cnidaria, sister to Cnidaria, Placozoa and Bilateria, and sister to all other animals. A series of studies that looked at the presence and absence of members of gene families and signalling pathways (e.g., homeoboxes, nuclear receptors, the Wnt signaling pathway, and sodium channels) showed evidence congruent with the latter two scenarios, that ctenophores are either sister to Cnidaria, Placozoa and Bilateria or sister to all other animal phyla. Several more recent studies comparing complete sequenced genomes of ctenophores with other sequenced animal genomes have also supported ctenophores as the sister lineage to all other animals. This position would suggest that neural and muscle cell types were either lost in major animal lineages (e.g., Porifera and Placozoa) or that they evolved independently in the ctenophore lineage. However, other researchers have argued that the placement of Ctenophora as sister to all other animals is a statistical anomaly caused by the high rate of evolution in ctenophore genomes, and that Porifera (sponges) is the earliest-diverging animal taxon instead. In agreement with the latter point, the analysis of a very large sequence alignment at the metazoan taxonomic scale (1,719 proteins totalizing ca. 400,000 amino acid positions) showed that ctenophores emerge as the second-earliest branching animal lineage, and sponges are sister-group to all other multicellular animals. Ctenophores and sponges are also the only known animal phyla that lack any true hox genes.

Relationships within Ctenophora

Since all modern ctenophores except the beroids have cydippid-like larvae, it has widely been assumed that their last common ancestor also resembled cydippids, having an egg-shaped body and a pair of retractable tentacles. Richard Harbison's purely morphological analysis in 1985 concluded that the cydippids are not monophyletic, in other words do not contain all and only the descendants of a single common ancestor that was itself a cydippid. Instead he found that various cydippid families were more similar to members of other ctenophore orders than to other cydippids. He also suggested that the last common ancestor of modern ctenophores was either cydippid-like or beroid-like. A molecular phylogeny analysis in 2001, using 26 species, including 4 recently discovered ones, confirmed that the cydippids are not monophyletic and concluded that the last common ancestor of modern ctenophores was cydippid-like. It also found that the genetic differences between these species were very small – so small that the relationships between the Lobata, Cestida and Thalassocalycida remained uncertain. This suggests that the last common ancestor of modern ctenophores was relatively recent, and perhaps survived the Cretaceous–Paleogene extinction event 65.5 million years ago while other lineages perished. When the analysis was broadened to include representatives of other phyla, it concluded that cnidarians are probably more closely related to bilaterians than either group is to ctenophores but that this diagnosis is uncertain.

Flatworm

The flatworms, flat worms, Platyhelminthes, Plathelminthes, or platyhelminths (from the Greek *platy*, meaning "flat" and (root), *helminth-*, meaning "worm") are a phylum of relatively simple bilaterian, unsegmented, soft-bodied invertebrates. Unlike other bilaterians, they are acoelomates (having no body cavity), and have no specialized circulatory and respiratory organs, which restricts them to having flattened shapes that allow oxygen and nutrients to pass through their bodies by diffusion. The digestive cavity has only one opening for both ingestion (intake of nutrients) and egestion (removal of undigested wastes); as a result, the food cannot be processed continuously.

In traditional medicinal texts, Platyhelminthes are divided into Turbellaria, which are mostly non-parasitic animals such as planarians, and three entirely parasitic groups: Cestoda, Trematoda and Monogenea; however, since the turbellarians have since been proven not to be monophyletic, this classification is now deprecated. Free-living flatworms are mostly predators, and live in water or in shaded, humid terrestrial environments, such as leaf litter. Cestodes (tapeworms) and trematodes (flukes) have complex life-cycles, with mature stages that live as parasites in the digestive systems of fish or land vertebrates, and intermediate stages that infest secondary hosts. The eggs of trematodes are excreted from their main hosts, whereas adult cestodes generate vast numbers of hermaphroditic, segment-like proglottids that detach when mature, are excreted, and then release eggs. Unlike the other parasitic groups, the monogeneans are external parasites infesting aquatic animals, and their larvae metamorphose into the adult form after attaching to a suitable host.

Because they do not have internal body cavities, Platyhelminthes were regarded as a primitive stage in the evolution of bilaterians (animals with bilateral symmetry and hence with distinct front and rear ends). However, analyses since the mid-1980s have separated out one subgroup, the Acoelomorpha, as basal bilaterians—closer to the original bilaterians than to any other modern groups. The remaining Platyhelminthes form a monophyletic group, one that contains all and only descendants of a common ancestor that is itself a member of the group. The redefined Platyhelminthes is part of the Lophotrochozoa, one of the three main groups of more complex bilaterians. These analyses had concluded the redefined Platyhelminthes, excluding Acoelomorpha, consists of two monophyletic subgroups, Catenulida and Rhabditophora, with Cestoda, Trematoda and Monogenea forming a monophyletic subgroup within one branch of the Rhabditophora. Hence, the traditional platyhelminth subgroup "Turbellaria" is now regarded as paraphyletic, since it excludes the wholly parasitic groups, although these are descended from one group of "turbellarians".

Over half of all known flatworm species are parasitic, and some do enormous harm to humans and their livestock. Schistosomiasis, caused by one genus of trematodes, is the second-most devastating of all human diseases caused by parasites, surpassed only by malaria. Neurocysticercosis, which arises when larvae of the pork tapeworm *Taenia solium* penetrate the central nervous system, is the major cause of acquired epilepsy worldwide. The threat of platyhelminth parasites to humans in developed countries is rising because of the popularity of raw or lightly cooked foods, and imports of food from high-risk areas. In less developed countries, people often cannot afford the fuel required to cook food thoroughly, and poorly designed water-supply and irrigation projects increase the dangers presented by poor sanitation and unhygienic farming.

Two planarian species have been used successfully in the Philippines, Indonesia, Hawaii, New Guinea, and Guam to control populations of the imported giant African snail *Achatina fulica*, which was displacing native snails. However, there is now concern that these planarians may themselves become a serious threat to native snails. In northwest Europe, there are concerns about the spread of the New Zealand planarian *Arthurdendyus triangulatus*, which preys on earthworms.

Description

Varied flatworm species from *Kunstformen der Natur (1904), plate 75*

Distinguishing Features

Platyhelminthes are bilaterally symmetrical animals: their left and right sides are mirror images of each other; this also implies they have distinct top and bottom surfaces and distinct head and tail ends. Like other bilaterians, they have three main cell layers (endoderm, mesoderm, and ectoderm), while the radially symmetrical cnidarians and ctenophores (comb jellies) have only two cell layers. Beyond that, they are "defined more by what they do not have than by any particular series of specializations." Unlike other bilaterians, Platyhelminthes have no internal body cavity, so are described as acoelomates. They also lack specialized circulatory and respiratory organs, both of these facts are defining features when classifying a flatworm's anatomy. Their bodies are soft and unsegmented.

	Cnidarians and Ctenophores	Platyhelminthes (flatworms)	More "advanced" bilaterians
Bilateral symmetry	No	Yes	
Number of main cell layers	Two, with jelly-like layer between them	Three	

Distinct brain	No	Yes
Specialized digestive system	No	Yes
Specialized excretory system	No	Yes
Body cavity containing internal organs	No	Yes
Specialized circulatory and respiratory organs	No	Yes

Features Common to all Subgroups

The lack of circulatory and respiratory organs limits platyhelminths to sizes and shapes that enable oxygen to reach and carbon dioxide to leave all parts of their bodies by simple diffusion. Hence, many are microscopic and the large species have flat ribbon-like or leaf-like shapes. The guts of large species have many branches, allowing nutrients to diffuse to all parts of the body. Respiration through the whole surface of the body makes them vulnerable to fluid loss, and restricts them to environments where dehydration is unlikely: sea and freshwater, moist terrestrial environments such as leaf litter or between grains of soil, and as parasites within other animals.

The space between the skin and gut is filled with mesenchyme, a connective tissue made of cells and reinforced by collagen fibers that act as a type of skeleton, providing attachment points for muscles. The mesenchyme contains all the internal organs and allows the passage of oxygen, nutrients and waste products. It consists of two main types of cell: fixed cells, some of which have fluid-filled vacuoles; and stem cells, which can transform into any other type of cell, and are used in regenerating tissues after injury or asexual reproduction.

Most platyhelminths have no anus and regurgitate undigested material through the mouth. However, some long species have an anus and some with complex, branched guts have more than one anus, since excretion only through the mouth would be difficult for them. The gut is lined with a single layer of endodermal cells that absorb and digest food. Some species break up and soften food first by secreting enzymes in the gut or pharynx (throat).

All animals need to keep the concentration of dissolved substances in their body fluids at a fairly constant level. Internal parasites and free-living marine animals live in environments with high concentrations of dissolved material, and generally let their tissues have the same level of concentration as the environment, while freshwater animals need to prevent their body fluids from becoming too dilute. Despite this difference in environments, most platyhelminths use the same system to control the concentration of their body fluids. Flame cells, so called because the beating of their flagella looks like a flickering candle flame, extract from the mesenchyme water that contains wastes and some reusable material, and drive it into networks of tube cells which are lined with flagella and microvilli. The tube cells' flagella drive the water towards exits called nephridiopores, while their microvilli reabsorb reusable materials and as much water as is needed to keep the body fluids at the right concentration. These combinations of flame cells and tube cells are called protonephridia.

In all platyhelminths, the nervous system is concentrated at the head end. This is least marked in the acoels, which have nerve nets rather like those of cnidarians and ctenophores, but densest around the head. Other platyhelminths have rings of ganglia in the head and main nerve trunks running along their bodies.

Major Subgroups

Early classification divided the flatworms into four groups: Turbellaria, Trematoda, Monogenea and Cestoda. This classification had long been recognized to be artificial, and in 1985, Ehlers proposed a phylogenetically more correct classification, where the massively polyphyletic "Turbellaria" was split into a dozen orders, and Trematoda, Monogenea and Cestoda were joined in the new order Neodermata. However, the classification presented here is the early, traditional, classification, as it still is the one used everywhere except in scientific articles.

Turbellaria

The turbellarian *Pseudoceros dimidiatus*

These have about 4,500 species, are mostly free-living, and range from 1 mm (0.039 in) to 600 mm (24 in) in length. Most are predators or scavengers, and terrestrial species are mostly nocturnal and live in shaded, humid locations, such as leaf litter or rotting wood. However, some are symbiotes of other animals, such as crustaceans, and some are parasites. Free-living turbellarians are mostly black, brown or gray, but some larger ones are brightly colored. The Acoela and Nemertodermatida were traditionally regarded as turbellarians, but are now regarded as members of a separate phylum, the Acoelomorpha, or as two separate phyla. *Xenoturbella*, a genus of very simple animals, has also been reclassified as a separate phylum.

Some turbellarians have a simple pharynx lined with cilia and generally feed by using cilia to sweep food particles and small prey into their mouths, which are usually in the middle of their undersides. Most other turbellarians have a pharynx that is eversible (can be extended by being turned inside-out), and the mouths of different species can be anywhere along the underside. The freshwater species *Microstomum caudatum* can open its mouth almost as wide as its body is long, to swallow prey about as large as itself.

Most turbellarians have pigment-cup ocelli ("little eyes"), one pair in most species, but two or even three pairs in some. A few large species have many eyes in clusters over the brain, mounted on tentacles, or spaced uniformly around the edge of the body. The ocelli can only distinguish the direction from which light is coming and enable the animals to avoid it. A few groups have statocysts, fluid-filled chambers containing a small, solid particle or, in a few groups, two. These statocysts are thought to be balance and acceleration sensors, as that is the function they perform in cnidarian medusae and in ctenophores. However, turbellarian statocysts have no sensory cilia, and how they sense the movements and positions of the solid particles is unknown. On the other hand, most have ciliated touch-sensor cells scattered over their bodies, especially on tentacles and around the edges. Specialized cells in pits or grooves on the head are probably smell sensors.

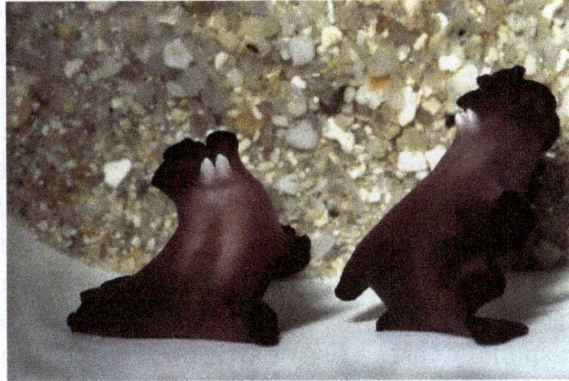

Two turbellarians are mating by penis fencing. Each has two penises, the white spikes on the undersides of their heads.

Planarians, a subgroup of seriates, are famous for their ability to regenerate if divided by cuts across their bodies. Experiments show that, in fragments that do not already have a head, a new head grows most quickly on those closest to the original head. This suggests the growth of a head is controlled by a chemical whose concentration diminishes from head to tail. Many turbellarians clone themselves by transverse or longitudinal division, and others, especially acoels, reproduce by budding.

The vast majority of turbellarians are hermaphrodites (have both female and male reproductive cells), and fertilize eggs internally by copulation. Some of the larger aquatic species mate by penis fencing, a duel in which each tries to impregnate the other, and the loser adopts the female role of developing the eggs. In most species, "miniature adults" emerge when the eggs hatch, but a few large species produce plankton-like larvae.

Trematoda

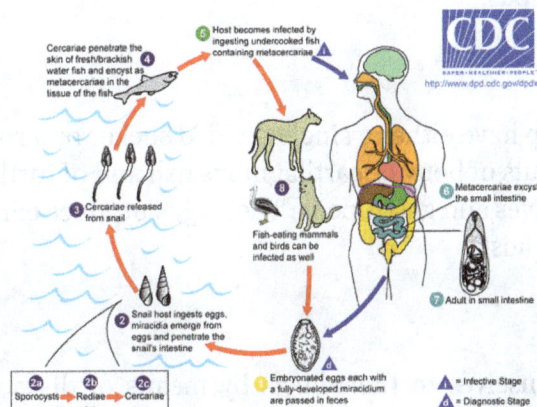

Life cycle of the diagenean *Metagonimus*

These parasites' name refers to the cavities in their holdfasts (hole), which resemble suckers and anchor them within their hosts. The skin of all species is a syncitium, a layer of cells that shares a single external membrane. Trematodes are divided into two groups, Digenea and Aspidogastrea (also known as Aspodibothrea).

Digenea

These are often called flukes, as most have flat rhomboid shapes like that of a flounder (Old English *flóc*). There are about 11,000 species, more than all other platyhelminthes combined, and second only to roundworms among parasites on metazoans. Adults usually have two holdfasts, a ring around the mouth and a larger sucker midway along what would be the underside in a free-living flatworm. Although the name "Digeneans" means "two generations", most have very complex life cycles with up to seven stages, depending on what combinations of environments the early stages encounter – most importantly whether the eggs are deposited on land or in water. The intermediate stages transfer the parasites from one host to another. The definitive host in which adults develop is a land vertebrate, the earliest host of juvenile stages is usually a snail that may live on land or in water, and in many cases a fish or arthropod is the second host. For example, the adjoining illustration shows the life cycle of the intestinal fluke *metagonimus*, which hatches in the intestine of a snail; moves to a fish, where it penetrates the body and encysts in the flesh; then moves to the small intestine of a land animal that eats the fish raw; and then generates eggs that are excreted and ingested by snails, thereby completing the cycle. Schistosomes, which cause the devastating tropical disease bilharzia, belong to this group.

Adults range between 0.2 mm (0.0079 in) and 6 mm (0.24 in) in length. Individual adult digeneans are of a single sex, and in some species, slender females live in enclosed grooves that run along the bodies of the males, and partially emerge to lay eggs. In all species, the adults have complex reproductive systems and can produce between 10,000 and 100,000 times as many eggs as a free-living flatworm. In addition, the intermediate stages that live in snails reproduce asexually.

Adults of different species infest different parts of the definitive host, for example the intestine, lungs, large blood vessels, and liver. The adults use a relatively large, muscular pharynx to ingest cells, cell fragments, mucus, body fluids or blood. In both the adults and the stages that live in snails, the external syncytium absorbs dissolved nutrients from the host. Adult digeneans can live without oxygen for long periods.

Aspidogastrea

Members of this small group have either a single divided sucker or a row of suckers that cover the underside. They infest the guts of bony or cartilaginous fish and of turtles, and the body cavities of marine and freshwater bivalves and gastropods. Their eggs produce ciliated swimming larvae, and the life cycle has one or two hosts.

Cercomeromorpha

These parasites attach themselves to their hosts by means of disks that bear crescent-shaped hooks. They are divided into Monogenea and Cestoda.

Monogenea

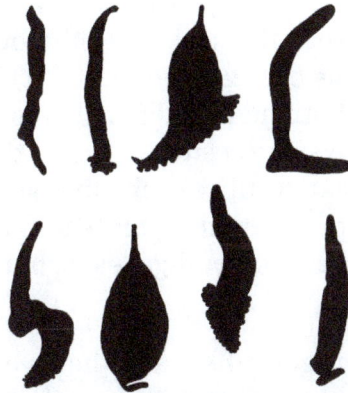

Silhouettes of bodies of various polyopisthocotylean Monogeneans

Of about 1,100 species of monogeneans, most are external parasites that require particular host species, mainly fish, but in some cases amphibians or aquatic reptiles. However, a few are internal parasites. Adult monogeneans have large attachment organs at the rear, haptors (*haptein*, means "catch"), which have suckers, clamps, and hooks. They often have flattened bodies. In some species, the pharynx secretes enzymes to digest the host's skin, allowing the parasite to feed on blood and cellular debris. Others graze externally on mucus and flakes of the hosts' skins. The name "Monogenea" is based on the fact that these parasites have only one nonlarval generation.

Cestoda

These are often called tapeworms because of their flat, slender but very long bodies – the name "cestode" is derived from the Latin word *cestus*, which means "tape". The adults of all 3,400 cestode species are internal parasites. Cestodes have no mouths or guts, and the syncitial skin absorbs nutrients – mainly carbohydrates and amino acids – from the host, and also disguises it chemically to avoid attacks by the host's immune system. Shortage of carbohydrates in the host's diet stunts the growth of the parasites and kills some. Their metabolisms generally use simple but inefficient chemical processes, and they compensate by consuming large amounts of food relative to their size.

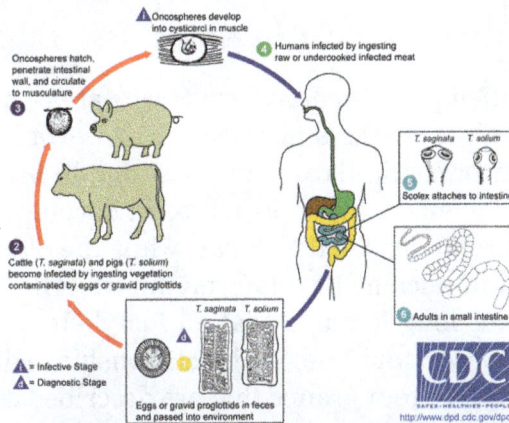

Life cycle of the eucestode *Taenia*: Inset 5 shows the scolex, which has four *Taenia solium*, a disk with hooks on the end. Inset 6 shows the tapeworm's whole body, in which the scolex is the tiny, round tip in the top left corner, and a mature proglottid has just detached.

In the majority of species, known as eucestodes ("true tapeworms"), the neck produces a chain of segments called proglottids by a process known as strobilation. Hence, the most mature proglottids are furthest from the scolex. Adults of *Taenia saginata*, which infests humans, can form proglottid chains over 20 metres (66 ft) long, although 4 metres (13 ft) is more typical. Each proglottid has both male and female reproductive organs. If the host's gut contains two or more adults of the same cestode species, they generally fertilize each other, but proglottids of the same worm can fertilize each other and even fertilize themselves. When the eggs are fully developed, the proglottids separate and are excreted by the host. The eucestode life cycle is less complex than that of digeneans, but varies depending on the species. For example:

- Adults of *Diphyllobothrium* infest fish, and the juveniles use copepod crustaceans as intermediate hosts. Excreted proglottids release their eggs into the water, and the eggs hatch into ciliated, swimming larvae. If a larva is swallowed by a copepod, it sheds the cilia and the skin becomes a syncitium and the larva makes its way into the copepod's hemocoel (internal cavity that is the main part of the circulatory system) and attached itself with three small hooks. If the copepod is eaten by a fish, the larva metamorphoses into a small, unsegmented tapeworm, drills through to the gut and becomes an adult.

- Various species of *Taenia* infest the guts of humans, cats and dogs. The juveniles use herbivores – for example pigs, cattle and rabbits – as intermediate hosts. Excreted proglottids release eggs that stick to grass leaves and hatch after being swallowed by a herbivore. The larva makes its way to the herbivore's muscles and metamorphoses into an oval worm about 10 millimetres (0.39 in) long, with a scolex that is kept inside. When the definitive host eats infested and raw or undercooked meat from an intermediate host, the worm's scolex pops out and attaches itself to the gut, and the adult tapeworm develops.

A members of the smaller group known as Cestodaria have no scolex, do not produce proglottids, and have body shapes like those of diageneans. Cestodarians parasitize fish and turtles.

Classification and Evolutionary Relationships

The oldest confidently identified parasitic flatworm fossils are cestode eggs found in a Permian shark coprolite, but helminth hooks still attached to Devonian acanthodians and placoderms might also represent parasitic flatworms with simple life cycles. The oldest known free-living platyhelminth specimen is a fossil preserved in Eocene age Baltic amber and placed in the monotypic species *Micropalaeosoma balticus*, while the oldest subfossil specimens are schistosome eggs discovered in ancient Egyptian mummies. The Platyhelminthes have very few synapomorphies, distinguishing features that all Platyhelminthes and no other animals have. This makes it difficult to work out both their relationships with other groups of animals and the relationships between different groups that are described as members of the Platyhelminthes.

The "traditional" view before the 1990s was that Platyhelminthes formed the sister group to all the other bilaterians, which include, for example, arthropods, molluscs, annelids and chordates.

Since then molecular phylogenetics, which aims to work out evolutionary "family trees" by comparing different organisms' biochemicals such as DNA, RNA and proteins, has radically changed scientists' view of evolutionary relationships between animals. Detailed morphological analyses of anatomical features in the mid-1980s and molecular phylogenetics analyses since 2000 using different sections of DNA agree that Acoelomorpha, consisting of Acoela (traditionally regarded as very simple "turbellarians") and Nemertodermatida (another small group previously classified as "turbellarians") are the sister group to all other bilaterians, including the rest of the Platyhelminthes. However, a 2007 study concluded that Acoela and Nemertodermatida were two distinct groups of bilaterians, although it agreed that both are more closely related to cnidarians (jellyfish, etc.) than other bilaterians are.

Xenoturbella, a bilaterian whose only well-defined organ is a statocyst, was originally classified as a "primitive turbellarian". However, it has recently been reclassified as a deuterostome.

The Platyhelminthes excluding Acoelomorpha contain two main groups, Catenulida and Rhabditophora, both of which are generally agreed to be monophyletic (each contains all and only the descendants of an ancestor that is a member of the same group). Early molecular phylogenetics analyses of the Catenulida and Rhabditophora left uncertainties about whether these could be combined in a single monophyletic group, but a study in 2008 concluded they could, therefore Platyhelminthes could be redefined as Catenulida plus Rhabditophora, excluding the Acoelomorpha.

Other molecular phylogenetics analyses agree the redefined Platyhelminthes are most closely related to Gastrotricha, and both are part of a grouping known as Platyzoa. Platyzoa are generally agreed to be at least closely related to the Lophotrochozoa, a superphylum that includes molluscs and annelid worms. The majority view is that Platyzoa are part of Lophotrochozoa, but a significant minority of researchers regard Platyzoa as a sister group of Lophotrochozoa.

It has been agreed since 1985 that each of the wholly parasitic platyhelminth groups (Cestoda, Monogenea and Trematoda) is monophyletic, and that together these form a larger monophyletic grouping, the Neodermata, in which the adults of all members have syncitial skins. However, there is debate about whether the Cestoda and Monogenea can be combined as an intermediate monophyletic group, the Cercomeromorpha, within the Neodermata. It is generally agreed that the Neodermata are a sub-group a few levels down in the "family tree" of the Rhabditophora. Hence the traditional sub-phylum "Turbellaria" is paraphyletic, since it does not include the Neodermata although these are descendants of a sub-group of "turbellarians".

Evolution

An outline of the origins of the parasitic life style has been proposed. Epithelial feeding monopisthocotyleans on fish hosts are basal in the Neodermata and were the first shift to parasitism from free living ancestors. The next evolutionary step was a dietary change from epithelium to blood. The last common ancestor of Digenea + Cestoda was monogenean and most likely sanguinivorous.

The earliest known fossils of tapeworms have been dated to 270 million years ago. They were found in coprolites (fossilised faeces) from an elasmobranch.

Interaction with Humans

Parasitism

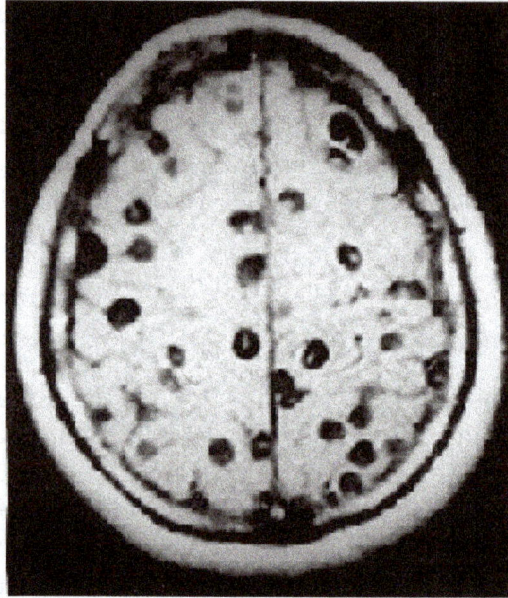

Magnetic resonance image of a patient with neurocysticercosis
demonstrating multiple cysticerci within the brain

Cestodes (tapeworms) and digeneans (flukes) cause important diseases in humans and their livestock, and monogeneans can cause serious losses of stocks in fish farms. Schistosomiasis, also known as bilharzia or snail fever, is the second-most devastating parasitic disease in tropical countries, behind malaria. The Carter Center estimated 200 million people in 74 countries are infected with the disease, and half the victims live in Africa. The condition has a low mortality rate, but often is a chronic illness that can damage internal organs. It can impair the growth and cognitive development of children, and increase the risk of bladder cancer in adults. The disease is caused by several flukes of the genus *Schistosoma*, which can bore through human skin. The people most at risk are those who use infected bodies of water for recreation or laundry.

In 2000, an estimated 45 million people were infected with the beef tapeworm *Taenia saginata* and 3 million with the pork tapeworm *Taenia solium*. Infection of the digestive system by adult tapeworms causes abdominal symptoms that are unpleasant but not disabling or life-threatening. However, neurocysticercosis resulting from penetration of *T. solium* larvae into the central nervous system is the major cause of acquired epilepsy worldwide. In 2000, about 39 million people were infected with trematodes (flukes) that naturally parasitize fish and crustaceans, but can pass to humans who eat raw or lightly cooked seafood. Infection of humans by the broad fish tapeworm *Diphyllobothrium latum* occasionally causes vitamin B_{12} deficiency and, in severe cases, megaloblastic anemia.

The threat to humans in developed countries is rising as a result of social trends: the increase in organic farming, which uses manure and sewage sludge rather than artificial fertilizers, and spreads parasites both directly and via the droppings of seagulls which feed on manure and sludge; the increasing popularity of raw or lightly cooked foods; imports of meat, seafood and salad vegetables from high-risk areas; and, as an underlying cause, reduced awareness of parasites compared with other public health issues such as pollution. In less-developed countries, inadequate sanitation and the use of human feces (night soil) as fertilizer and to enrich fish farm ponds continues to spread parasitic platyhelminths, and poorly designed water-supply and irrigation projects have provided additional channels for their spread. People in these countries often cannot afford the cost of fuel required to cook food thoroughly enough to kill parasites. Controlling parasites that infect humans and livestock has become more difficult, as many species have become resistant to drugs that used to be effective, mainly for killing juveniles in meat. While poorer countries still struggle with unintentional infection, cases have been reported of intentional infection in the US by dieters desperate for rapid weight-loss.

Pests

There is concern about the proliferation in northwest Europe, including the British Isles, of the New Zealand planarian *Arthurdendyus triangulatus* and the Australian flatworm *Australoplana sanguinea*, both of which prey on earthworms. *A. triangulatus* is thought to have reached Europe in containers of plants imported by botanical gardens.

Benefits

In Hawaii, the planarian *Endeavouria septemlineata* has been used to control the imported giant African snail *Achatina fulica*, which was displacing native snails, and *Platydemus manokwari*, another planarian, has been used for the same purpose in Philippines, Indonesia, New Guinea and Guam. Although *A. fulica* has declined sharply in Hawaii, there are doubts about how much *E. septemlineata* contributed to this. However, *P. manokwari* is given credit for severely reducing, and in places exterminating, *A. fulica* – achieving much greater success than most biological pest control programs, which generally aim for a low, stable population of the pest species. The ability of planarians to take different kinds of prey and to resist starvation may account for their ability to decimate *A. fulica*. However, these abilities have raised concerns that planarians may themselves become a serious threat to native snails.

A study in La Plata, Argentina shows the potential for planarians such as *Girardia anceps*, *Mesostoma ehrenbergii*, and *Bothromesostoma evelinae* to reduce populations of the mosquito species *Aedes aegypti* and *Culex pipiens*. The experiment showed that *G. anceps* in particular can prey on all instars of both mosquito species and maintain a steady predation rate over time. The ability for these flatworms to live in artificial containers showed the potential of placing these species in popular mosquito breeding sites, which would ideally reduce the amount of mosquito-borne diseases

Annelid

The annelids (*Annelida*, from Latin *anellus*, "little ring"), also known as the ringed worms or segmented worms, are a large phylum, with over 17,000 extant species including ragworms, earthworms, and leeches. The species exist in and have adapted to various ecologies - some in marine environments as distinct as tidal zones and hydrothermal vents, others in fresh water, and yet others in moist terrestrial environments.

The annelids are bilaterally symmetrical, triploblastic, coelomate, invertebrate organisms. They also have parapodia for locomotion. Most textbooks still use the traditional division into polychaetes (almost all marine), oligochaetes (which include earthworms) and leech-like species. Cladistic research since 1997 has radically changed this scheme, viewing leeches as a sub-group of oligochaetes and oligochaetes as a sub-group of polychaetes. In addition, the Pogonophora, Echiura and Sipuncula, previously regarded as separate phyla, are now regarded as sub-groups of polychaetes. Annelids are considered members of the Lophotrochozoa, a "super-phylum" of protostomes that also includes molluscs, brachiopods, flatworms and nemerteans.

The basic annelid form consists of multiple segments. Each segment has the same sets of organs and, in most polychates, has a pair of parapodia that many species use for locomotion. Septa separate the segments of many species, but are poorly defined or absent in others, and Echiura and Sipuncula show no obvious signs of segmentation. In species with well-developed septa, the blood circulates entirely within blood vessels, and the vessels in segments near the front ends of these species are often built up with muscles that act as hearts. The septa of such species also enable them to change the shapes of individual segments, which facilitates movement by peristalsis ("ripples" that pass along the body) or by undulations that improve the effectiveness of the parapodia. In species with incomplete septa or none, the blood circulates through the main body cavity without any kind of pump, and there is a wide range of locomotory techniques – some burrowing species turn their pharynges inside out to drag themselves through the sediment.

Although many species can reproduce asexually and use similar mechanisms to regenerate after severe injuries, sexual reproduction is the normal method in species whose reproduction has been studied. The minority of living polychaetes whose reproduction and lifecycles are known produce trochophore larvae, that live as plankton and then sink and metamorphose into miniature adults. Oligochaetes are full hermaphrodites and produce a ring-like cocoon around their bodies, in which the eggs and hatchlings are nourished until they are ready to emerge.

Earthworms are Oligochaetes that support terrestrial food chains both as prey and in some regions are important in aeration and enriching of soil. The burrowing of marine polychaetes, which may constitute up to a third of all species in near-shore environments, encourages the development of ecosystems by enabling water and oxygen to penetrate the sea floor. In addition to improving soil fertility, annelids serve humans as food and as bait. Scientists observe annelids to monitor the quality of marine and fresh water. Although blood-letting is no longer in favor with doctors, some leech species are regarded as endangered species because they have been over-harvested for this purpose in the last few centuries. Ragworms' jaws are now being studied by engineers as they offer an exceptional combination of lightness and strength.

Since annelids are soft-bodied, their fossils are rare – mostly jaws and the mineralized tubes that some of the species secreted. Although some late Ediacaran fossils may represent annelids, the oldest known fossil that is identified with confidence comes from about 518 million years ago in the early Cambrian period. Fossils of most modern mobile polychaete groups appeared by the end of the Carboniferous, about 299 million years ago. Palaeontologists disagree about whether some body fossils from the mid

Ordovician, about 472 to 461 million years ago, are the remains of oligochaetes, and the earliest indisputable fossils of the group appear in the Tertiary period, which began 65 million years ago.

Classification and Diversity

There are over 22,000 living annelid species, ranging in size from microscopic to the Australian giant Gippsland earthworm and *Amynthas mekongianus* (Cognetti, 1922), which can both grow up to 3 metres (9.8 ft) long. Although research since 1997 has radically changed scientists' views about the evolutionary family tree of the annelids, most textbooks use the traditional classification into the following sub-groups:

- Polychaetes (about 12,000 species). As their name suggests, they have multiple chetae ("hairs") per segment. Polychaetes have parapodia that function as limbs, and nuchal organs that are thought to be chemosensors. Most are marine animals, although a few species live in fresh water and even fewer on land.

An earthworm's clitellum

- Clitellates (about 10,000 species). These have few or no chetae per segment, and no nuchal organs or parapodia. However, they have a unique reproductive organ, the ring-shaped clitellum ("pack saddle") around their bodies, which produces a cocoon that stores and nourishes fertilized eggs until they hatch or, in moniligastrids, yolky eggs that provide nutrition for the embyros. The clitellates are sub-divided into:

- Oligochaetes ("with few hairs"), which includes earthworms. Oligochaetes have a sticky pad in the roof of the mouth. Most are burrowers that feed on wholly or partly decomposed organic materials.

- Hirudinea, whose name means "leech-shaped" and whose best known members are leeches. Marine species are mostly blood-sucking parasites, mainly on fish, while most freshwater species are predators. They have suckers at both ends of their bodies, and use these to move rather like inchworms.

The Archiannelida, minute annelids that live in the spaces between grains of marine sediment, were treated as a separate class because of their simple body structure, but are now regarded as polychaetes. Some other groups of animals have been classified in various ways, but are now widely regarded as annelids:

- Pogonophora / Siboglinidae were first discovered in 1914, and their lack of a recognizable gut made it difficult to classify them. They have been classified as a separate phylum, Pogonophora, or as two phyla, Pogonophora and Vestimentifera. More recently they have been re-classified as a family, Siboglinidae, within the polychaetes.

- The Echiura have a checkered taxonomic history: in the 19th century they were assigned to the phylum "Gephyrea", which is now empty as its members have been assigned to other phyla; the Echiura were next regarded as annelids until the 1940s, when they were classified as a phylum in their own right; but a molecular phylogenetics analysis in 1997 concluded that echiurans are annelids.

- Myzostomida live on crinoids and other echinoderms, mainly as parasites. In the past they have been regarded as close relatives of the trematode flatworms or of the tardigrades, but in 1998 it was suggested that they are a sub-group of polychaetes. However, another analysis in 2002 suggested that myzostomids are more closely related to flatworms or to rotifers and acanthocephales.

Distinguishing Features

No single feature distinguishes Annelids from other invertebrate phyla, but they have a distinctive combination of features. Their bodies are long, with segments that are divided externally by shallow ring-like constrictions called annuli and internally by septa ("partitions") at the same points, although in some species the septa are incomplete and in a few cases missing. Most of the segments contain the same sets of organs, although sharing a common gut, circulatory system and nervous system makes them inter-dependent. Their bodies are covered by a cuticle (outer covering) that does not contain cells but is secreted by cells in the skin underneath, is made of tough but flexible collagen and does not molt – on the other hand arthropods' cuticles are made of the more rigid α-chitin, and molt until the arthropods reach their full size. Most annelids have closed circulatory systems, where the blood makes its entire circuit via blood vessels.

Distinguishing Features						
	Annelida	**Recently merged into Annelida**		**Closely related**	**Similar-looking phyla**	
		Echiura	**Sipuncula**	**Nemertea**	**Arthropoda**	**Onychophora**
External segmentation	Yes	no	no	Only in a few species	Yes, except in mites	no
Repetition of internal organs	Yes	no	no	Yes	In primitive forms	Yes
Septa between segments	In most species	no	no	No	No	No

Cuticle material	collagen	collagen	collagen	none	α-chitin	α-chitin
Molting	Generally no; but some polychaetes molt their jaws, and leeches molt their skins	no	no	no	Yes	Yes
Body cavity	Coelom; but this is reduced or missing in many leeches and some small polychaetes	2 coelomata, main and in proboscis	2 coelomata, main and in tentacles	Coelom only in proboscis	Hemocoel	Hemocoel
Circulatory system	Closed in most species	Open outflow, return via branched vein	Open	Closed	Open	Open

Description

Segmentation

Most of an annelid's body consists of segments that are practically identical, having the same sets of internal organs and external chaetae and, in some species, appendages. However, the frontmost and rearmost sections are not regarded as true segments as they do not contain the standard sets of organs and do not develop in the same way as the true segments. The frontmost section, called the prostomium contains the brain and sense organs, while the rearmost, called the pygidium or periproct contains the anus, generally on the underside. The first section behind the prostomium, called the peristomium is regarded by some zoologists as not a true segment, but in some polychaetes the peristomium has chetae and appendages like those of other segments.

The segments develop one at a time from a growth zone just ahead of the pygidium, so that an annelid's youngest segment is just in front of the growth zone while the peristomium is the oldest. This pattern is called teloblastic growth. Some groups of annelids, including all leeches, have fixed maximum numbers of segments, while others add segments throughout their lives.

The phylum's name is derived from the Latin word *annelus*, meaning "little ring".

Body Wall, Chetae and Parapodia

Annelids' cuticles are made of collagen fibers, usually in layers that spiral in alternating directions so that the fibers cross each other. These are secreted by the one-cell deep epidermis (outermost skin layer). A few marine annelids that live in tubes lack cuticles, but their tubes have a similar structure, and mucus-secreting glands in the epidermis protect their skins. Under the epidermis is the dermis, which is made of connective tissue, in other words a combination of cells and non-cellular materials such as collagen. Below this are two layers of muscles, which develop from the lining of the coelom (body cavity): circular muscles make a segment longer and slimmer when they contract, while under them are longitudinal muscles, usually four distinct strips, whose contractions make the segment shorter and fatter. Some annelids also have oblique internal muscles that connect the underside of the body to each side.

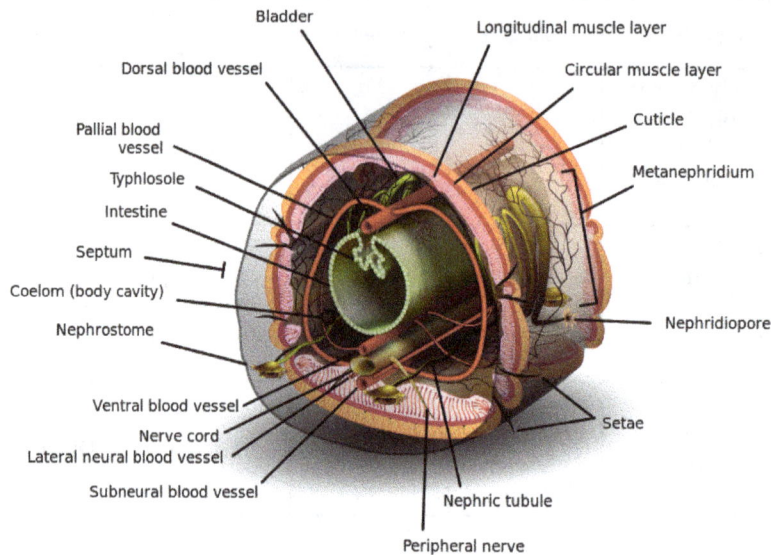

The setae ("hairs") of annelids project out from the epidermis to provide traction and other capabilities. The simplest are unjointed and form paired bundles near the top and bottom of each side of each segment. The parapodia ("limbs") of annelids that have them often bear more complex chetae at their tips – for example jointed, comb-like or hooked. Chetae are made of moderately flexible β-chitin and are formed by follicles, each of which has a chetoblast ("hair-forming") cell at the bottom and muscles that can extend or retract the cheta. The chetoblasts produce chetae by forming microvilli, fine hair-like extensions that increase the area available for secreting the cheta. When the cheta is complete, the microvilli withdraw into the chetoblast, leaving parallel tunnels that run almost the full length of the cheta. Hence annelids' chetae are structurally different from the setae ("bristles") of arthropods, which are made of the more rigid α-chitin, have a single internal cavity, and are mounted on flexible joints in shallow pits in the cuticle.

Nearly all polychaetes have parapodia that function as limbs, while other major annelid groups lack them. Parapodia are unjointed paired extensions of the body wall, and their muscles are derived from the circular muscles of the body. They are often supported internally by one or more large, thick chetae. The parapodia of burrowing and tube-dwelling polychaetes are often just ridges whose tips bear hooked chetae. In active crawlers and swimmers the parapodia are often divided into large upper and lower paddles on a very short trunk, and the paddles are generally fringed with chetae and sometimes with cirri (fused bundles of cilia) and gills.

Nervous System and Senses

The brain generally forms a ring round the pharynx (throat), consisting of a pair of ganglia (local control centers) above and in front of the pharynx, linked by nerve cords either side of the pharynx to another pair of ganglia just below and behind it. The brains of polychaetes are generally in the prostomium, while those of clitellates are in the peristomium or sometimes the first segment behind the peristomium. In some very mobile and active polychaetes the brain is enlarged and more complex, with visible hindbrain, midbrain and forebrain sections. The rest of the central nervous system is generally "ladder-like", consisting of a pair of nerve cords that run through the bottom part of the body and have in each segment paired ganglia linked by a transverse connection. From each segmental ganglion a branching system of local nerves runs into the body wall and then encircles the body. However, in most polychaetes the two main nerve cords are fused, and in the tube-dwelling genus *Owenia* the single nerve chord has no ganglia and is located in the epidermis.

As in arthropods, each muscle fiber (cell) is controlled by more than one neuron, and the speed and power of the fiber's contractions depends on the combined effects of all its neurons. Vertebrates have a different system, in which one neuron controls a group of muscle fibers. Most annelids' longitudinal nerve trunks include giant axons (the output signal lines of nerve cells). Their large diameter decreases their resistance, which allows them to transmit signals exceptionally fast. This enables these worms to withdraw rapidly from danger by shortening their bodies. Experiments have shown that cutting the giant axons prevents this escape response but does not affect normal movement.

The sensors are primarily single cells that detect light, chemicals, pressure waves and contact, and are present on the head, appendages (if any) and other parts of the body. Nuchal ("on the neck") organs are paired, ciliated structures found only in polychaetes, and are thought to be chemosensors. Some polychaetes also have various combinations of ocelli ("little eyes") that detect the direction from which light is coming and camera eyes or compound eyes that can probably form images. The compound eyes probably evolved independently of arthropods' eyes. Some tube-worms use ocelli widely spread over their bodies to detect the shadows of fish, so that they can quickly withdraw into their tubes. Some burrowing and tube-dwelling polychaetes have statocysts (tilt and balance sensors) that tell them which way is down. A few polychaete genera have on the undersides of their heads palps that are used both in feeding and as "feelers", and some of these also have antennae that are structurally similar but probably are used mainly as "feelers".

Coelom, Locomotion and Circulatory System

Most annelids have a pair of coelomata (body cavities) in each segment, separated from other segments by septa and from each other by vertical mesenteries. Each septum forms a sandwich with connective tissue in the middle and mesothelium (membrane that serves as a lining) from the preceding and following segments on either side. Each mesentery is similar except that the mesothelium is the lining of each of the pair of coelomata, and the blood vessels and, in polychaetes, the main nerve cords are embedded in it. The mesothelium is made of modified epitheliomuscular cells; in other words, their bodies form part of the epithelium but their bases extend to form muscle fibers in the body wall. The mesothelium may also form radial and circular muscles on the septa, and circular muscles around the blood vessels and gut. Parts of the mesothelium, especially on the outside of the gut, may also form chloragogen cells that perform similar functions to the livers of vertebrates: producing and storing glycogen and fat; producing the oxygen-carrier hemoglobin; breaking down proteins; and turning nitrogenous waste products into ammonia and urea to be excreted.

Many annelids move by peristalsis (waves of contraction and expansion that sweep along the body), or flex the body while using parapodia to crawl or swim. In these animals the septa enable the circular and longitudinal muscles to change the shape of individual segments, by making each segment a separate fluid-filled "balloon". However, the septa are often incomplete in annelids that are semi-sessile or that do not move by peristalsis or by movements of parapodia – for example some move by whipping movements of the body, some small marine species move by means of cilia (fine muscle-powered hairs) and some burrowers turn their pharynges (throats) inside out to penetrate the sea-floor and drag themselves into it.

The fluid in the coelomata contains coelomocyte cells that defend the animals against parasites and infections. In some species coelomocytes may also contain a respiratory pigment – red hemoglobin in some species, green chlorocruorin in others (dissolved in the plasma) – and provide oxygen transport within their segments. Respiratory pigment is also dissolved in the blood plasma. Species with well-developed septa generally also have blood vessels running all long their bodies above and below the gut, the upper one carrying blood forwards while the lower one carries it backwards. Networks of capillaries in the body wall and around the gut transfer blood between the main blood vessels and to parts of the segment that need oxygen and nutrients. Both of the major vessels, especially the upper one, can pump blood by contracting. In some annelids the forward end of the upper blood vessel is enlarged with muscles to form a heart, while in the forward ends of many earthworms some of the vessels that connect the upper and lower main vessels function as hearts. Species with poorly developed or no septa generally have no blood vessels and rely on the circulation within the coelom for delivering nutrients and oxygen.

However, leeches and their closest relatives have a body structure that is very uniform within the group but significantly different from that of other annelids, including other members of the Clitellata. In leeches there are no septa, the connective tissue layer of the body wall is so thick that it occupies much of the body, and the two coelomata are widely separated and run the length of the body. They function as the main blood vessels, although they are side-by-side rather than upper and lower. However, they are lined with mesothelium, like the coelomata and unlike the blood vessels of other annelids. Leeches generally use suckers at their front and rear ends to move like inchworms. The anus is on the upper surface of the pygidium.

Respiration

In some annelids, including earthworms, all respiration is via the skin. However, many polychaetes and some clitellates (the group to which earthworms belong) have gills associated with most segments, often as extensions of the parapodia in polychaetes. The gills of tube-dwellers and burrowers usually cluster around whichever end has the stronger water flow.

Feeding and Excretion

Feeding structures in the mouth region vary widely, and have little correlation with the animals' diets. Many polychaetes have a muscular pharynx that can be everted (turned inside out to extend it). In these animals the foremost few segments often lack septa so that, when the muscles in these segments contract, the sharp increase in fluid pressure from all these segments everts the pharynx very quickly. Two families, the Eunicidae and Phyllodocidae, have evolved jaws, which can be used for seizing prey, biting off pieces of vegetation, or grasping dead and decaying matter. On the other hand, some predatory polychaetes have neither jaws nor eversible pharynges. Selective deposit feeders generally live in tubes on the sea-floor and use palps to find food particles in the sediment and then wipe them into their mouths. Filter feeders use "crowns" of palps covered in cilia that wash food particles towards their mouths. Non-selective deposit feeders ingest soil or marine sediments via mouths that are generally unspecialized. Some clitellates have sticky pads in the roofs of their mouths, and some of these can evert the pads to capture prey. Leeches often have an eversible proboscis, or a muscular pharynx with two or three teeth.

Lamellibrachian tube worms have no gut and gain nutrients from chemoautotrophic bacteria living inside them.

The gut is generally an almost straight tube supported by the mesenteries (vertical partitions within segments), and ends with the anus on the underside of the pygidium. However, in members of the tube-dwelling family Siboglinidae the gut is blocked by a swollen lining that houses symbiotic bacteria, which can make up 15% of the worms' total weight. The bacteria convert inorganic matter – such as hydrogen sulfide and carbon dioxide from hydrothermal vents, or methane from seeps – to organic matter that feeds themselves and their hosts, while the worms extend their palps into the gas flows to absorb the gases needed by the bacteria.

Annelids with blood vessels use metanephridia to remove soluble waste products, while those without use protonephridia. Both of these systems use a two-stage filtration process, in which fluid and waste products are first extracted and these are filtered again to re-absorb any re-usable materials while dumping toxic and spent materials as urine. The difference is that protonephridia combine both filtration stages in the same organ, while metanephridia perform only the second filtration and rely on other mechanisms for the first – in annelids special filter cells in the walls of the blood vessels let fluids and other small molecules pass into the coelomic fluid, where it circulates to the metanephridia. In annelids the points at which fluid enters the protonephridia or metanephridia are on the forward side of a septum while the second-stage filter and the nephridiopore (exit opening in the body wall) are in the following segment. As a result, the hindmost segment (before the growth zone and pygidium) has no structure that extracts its wastes, as there is no following segment to filter and discharge them, while the first segment contains an extraction structure that passes wastes to the second, but does not contain the structures that re-filter and discharge urine.

Reproduction and Life Cycle

Asexual Reproduction

Polychaetes can reproduce asexually, by dividing into two or more pieces or by budding off a new individual while the parent remains a complete organism. Some oligochaetes, such as *Aulophorus furcatus*, seem to reproduce entirely asexually, while others reproduce asexually in summer and sexually in autumn. Asexual reproduction in oligochaetes is always by dividing into two or more pieces, rather than by budding. However, leeches have never been seen reproducing asexually.

This sabellid tubeworm is budding

Most polychaetes and oligochaetes also use similar mechanisms to regenerate after suffering damage. Two polychaete genera, *Chaetopterus* and *Dodecaceria*, can regenerate from a single segment, and others can regenerate even if their heads are removed. Annelids are the most complex animals that can regenerate after such severe damage. On the other hand, leeches cannot regenerate.

Sexual Reproduction

It is thought that annelids were originally animals with two separate sexes, which released ova and sperm into the water via their nephridia. The fertilized eggs develop into trochophore larvae, which live as plankton. Later they sink to the sea-floor and metamorphose into miniature adults: the part of the trochophore between the apical tuft and the prototroch becomes the prostomium (head); a small area round the trochophore's anus becomes the pygidium (tail-piece); a narrow band immediately in front of that becomes the growth zone that produces new segments; and the rest of the trochophore becomes the peristomium (the segment that contains the mouth).

However, the lifecycles of most living polychaetes, which are almost all marine animals, are unknown, and only about 25% of the 300+ species whose lifecycles are known follow this pattern. About 14% use a similar external fertilization but produce yolk-rich eggs, which reduce the time the larva needs to spend among the plankton, or eggs from which miniature adults emerge rather than larvae. The rest care for the fertilized eggs until they hatch – some by producing jelly-covered masses of eggs which they tend, some by attaching the eggs to their bodies and a few species by keeping the eggs within their bodies until they hatch. These species use a variety of methods for sperm transfer; for example, in some the females collect sperm released into the water, while in others the males have a penis that inject sperm into the female. There is no guarantee that this is a representative sample of polychaetes' reproductive patterns, and it simply reflects scientists' current knowledge.

Some polychaetes breed only once in their lives, while others breed almost continuously or through several breeding seasons. While most polychaetes remain of one sex all their lives, a significant percentage of species are full hermaphrodites or change sex during their lives. Most polychaetes whose reproduction has been studied lack permanent gonads, and it is uncertain how they produce ova and sperm. In a few species the rear of the body splits off and becomes a separate individual that lives just long enough to swim to a suitable environment, usually near the surface, and spawn.

Most mature clitellates (the group that includes earthworms and leeches) are full hermaphrodites, although in a few leech species younger adults function as males and become female at maturity. All have well-developed gonads, and all copulate. Earthworms store their partners' sperm in spermathecae ("sperm stores") and then the clitellum produces a cocoon that collects ova from the ovaries and then sperm from the spermathecae. Fertilization and development of earthworm eggs takes place in the cocoon. Leeches' eggs are fertilized in the ovaries, and then transferred to the cocoon. In all clitellates the cocoon also either produces yolk when the eggs are fertilized or nutrients while they are developing. All clitellates hatch as miniature adults rather than larvae.

Ecological Significance

Charles Darwin's book *The Formation of Vegetable Mould through the Action of Worms* (1881) presented the first scientific analysis of earthworms' contributions to soil fertility. Some burrow while others live entirely on the surface, generally in moist leaf litter. The burrowers loosen the soil so that oxygen and water can penetrate it, and both surface and burrowing worms help to produce soil by mixing organic and mineral matter, by accelerating the decomposition of organic matter and thus making it more quickly available to other organisms, and by concentrating minerals and converting them to forms that plants can use more easily. Earthworms are also important prey for birds ranging in size from robins to storks, and for mammals ranging from shrews to badgers, and in some cases conserving earthworms may be essential for conserving endangered birds.

Terrestrial annelids can be invasive in some situations. In the glaciated areas of North America, for example, almost all native earthworms are thought to have been killed by the glaciers and the worms currently found in those areas are all introduced from other areas, primarily from Europe, and, more recently, from Asia. Northern hardwood forests are especially negatively impacted by invasive worms through the loss of leaf duff, soil fertility, changes in soil chemistry and the loss of ecological diversity. Especially of concern is *Amynthas agrestis* and at least one state (Wisconsin) has listed it as a prohibited species.

Earthworms migrate only a limited distance annually on their own, and the spread of invasive worms is increased rapidly by anglers and from worms or their cocoons in the dirt on vehicle tires or footwear.

Marine annelids may account for over one-third of bottom-dwelling animal species around coral reefs and in tidal zones. Burrowing species increase the penetration of water and oxygen into the sea-floor sediment, which encourages the growth of populations of aerobic bacteria and small animals alongside their burrows.

Although blood-sucking leeches do little direct harm to their victims, some transmit flagellates that can be very dangerous to their hosts. Some small tube-dwelling oligochaetes transmit myxosporean parasites that cause whirling disease in fish.

Interaction with Humans

Earthworms make a significant contribution to soil fertility. The rear end of the Palolo worm, a marine polychaete that tunnels through coral, detaches in order to spawn at the surface, and the people of Samoa regard these spawning modules as a delicacy. Anglers sometimes find that worms are more effective bait than artificial flies, and worms can be kept for several days in a tin lined with damp moss. Ragworms are commercially important as bait and as food sources for aquaculture, and there have been proposals to farm them in order to reduce over-fishing of their natural populations. Some marine polychaetes' predation on molluscs causes serious losses to fishery and aquaculture operations.

Scientists study aquatic annelids to monitor the oxygen content, salinity and pollution levels in fresh and marine water.

Accounts of the use of leeches for the medically dubious practise of blood-letting have come from China around 30 AD, India around 200 AD, ancient Rome around 50 AD and later throughout Europe. In the 19th century medical demand for leeches was so high that some areas' stocks were exhausted and other regions imposed restrictions or bans on exports, and *Hirudo medicinalis* is treated as an endangered species by both IUCN and CITES. More recently leeches have been used to assist in microsurgery, and their saliva has provided anti-inflammatory compounds and several important anticoagulants, one of which also prevents tumors from spreading.

Ragworms' jaws are strong but much lighter than the hard parts of many other organisms, which are biomineralized with calcium salts. These advantages have attracted the attention of engineers. Investigations showed that ragworm jaws are made of unusual proteins that bind strongly to zinc.

Evolutionary History

Fossil Record

Since annelids are soft-bodied, their fossils are rare. Polychaetes' fossil record consists mainly of the jaws that some species had and the mineralized tubes that some secreted. Some Ediacaran fossils such as *Dickinsonia* in some ways resemble polychaetes, but the similarities are too vague for these fossils to be classified with confidence. The small shelly fossil *Cloudina*, from 549 to 542 million years ago, has been classified by some authors as an annelid, but by others as a cnidarian (i.e. in the phylum to which jellyfish and sea anemones belong). Until 2008 the earliest fossils widely accepted as annelids were the polychaetes *Canadia* and *Burgessochaeta*, both from Canada's Burgess Shale, formed about 505 million years ago in the early Cambrian. *Myoscolex*, found in Australia and a little older than the Burgess Shale, was possibly an annelid. However, it lacks some typical annelid features and has features which are not usually found in annelids and some of which are associated with other phyla. Then Simon Conway Morris and John Peel reported *Phragmochaeta* from Sirius Passet, about 518 million years old, and concluded that it was the oldest annelid known to date. There has been vigorous debate about whether the Burgess Shale fossil *Wiwaxia* was a mollusc or an annelid. Polychaetes diversified in the early Ordovician, about 488 to 474 million years ago. It is not until the early Ordovician that the first annelid jaws are found, thus the crown-group cannot have appeared before this date and probably appeared somewhat later. By the end of the Carboniferous, about 299 million years ago, fossils of most of the modern mobile polychaete groups had appeared. Many fossil tubes look like those made by modern sessile polychaetes, but the first tubes clearly produced by polychaetes date from the Jurassic, less than 199 million years ago.

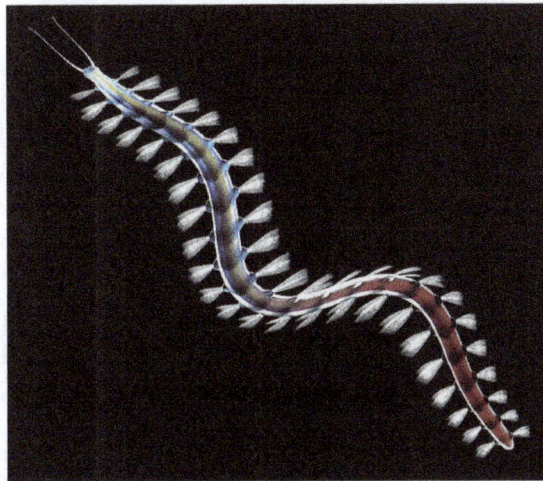

Burgessochaeta setigera

The earliest good evidence for oligochaetes occurs in the Tertiary period, which began 65 million years ago, and it has been suggested that these animals evolved around the same time as flowering plants in the early Cretaceous, from 130 to 90 million years ago. A trace fossil consisting of a convoluted burrow partly filled with small fecal pellets may be evidence that earthworms were present in the early Triassic period from 251 to 245 million years ago. Body fossils going back to the mid Ordovician, from 472 to 461 million years ago, have been tentatively classified as oligochaetes, but these identifications are uncertain and some have been disputed.

Family Tree

Traditionally the annelids have been divided into two major groups, the polychaetes and clitellates. In turn the clitellates were divided into oligochaetes, which include earthworms, and hirudinomorphs, whose best-known members are leeches. For many years there was no clear arrangement of the approximately 80 polychaete families into higher-level groups. In 1997 Greg Rouse and Kristian Fauchald attempted a "first heuristic step in terms of bringing polychaete systematics to an acceptable level of rigour", based on anatomical structures, and divided polychaetes into:

- Scolecida, less than 1,000 burrowing species that look rather like earthworms.

- Palpata, the great majority of polychaetes, divided into:

- Canalipalpata, which are distinguished by having long grooved palps that they use for feeding, and most of which live in tubes.

- Aciculata, the most active polychaetes, which have parapodia reinforced by internal spines (aciculae).

Also in 1997 Damhnait McHugh, using molecular phylogenetics to compare similarities and differences in one gene, presented a very different view, in which: the clitellates were an offshoot of one branch of the polychaete family tree; the pogonophorans and echiurans, which for a few decades had been regarded as a separate phyla, were placed on other branches of the polychaete tree. Subsequent molecular phylogenetics analyses on a similar scale presented similar conclusions.

In 2007 Torsten Struck and colleagues compared 3 genes in 81 taxa, of which 9 were outgroups, in other words not considered closely related to annelids but included to give an indication of where the organisms under study are placed on the larger tree of life. For a cross-check the study used an analysis of 11 genes (including the original 3) in 10 taxa. This analysis agreed that clitellates, pogonophorans and echiurans were on various branches of the polychaete family tree. It also concluded that the classification of polychaetes into Scolecida, Canalipalpata and Aciculata was useless, as the members of these alleged groups were scattered all over the family tree derived from comparing the 81 taxa. In addition, it also placed sipunculans, generally regarded at the time as a separate phylum, on another branch of the polychaete tree, and concluded that leeches were a sub-group of oligochaetes rather than their sister-group among the clitellates. Rouse accepted the analyses based on molecular phylogenetics, and their main conclusions are now the scientific consensus, although the details of the annelid family tree remain uncertain.

In addition to re-writing the classification of annelids and 3 previously independent phyla, the molecular phylogenetics analyses undermine the emphasis that decades of previous writings placed on the importance of segmentation in the classification of invertebrates. Polychaetes, which these analyses found to be the parent group, have completely segmented bodies, while polychaetes' echiurans and sipunculan offshoots are not segmented and pogonophores are segmented only in the rear parts of their bodies. It now seems that segmentation can appear and disappear much more easily in the course of evolution than was previously thought. The 2007 study also noted that the ladder-like nervous system, which is associated with segmentation, is less universal than previously thought in both annelids and arthropods.

Annelids are members of the protostomes, one of the two major superphyla of bilaterian animals – the other is the deuterostomes, which includes vertebrates. Within the protostomes, annelids used to be grouped with arthropods under the super-group Articulata ("jointed animals"), as segmentation is obvious in most members of both phyla. However, the genes that drive segmentation in arthropods do not appear to do the same in annelids. Arthropods and annelids both have close relatives that are unsegmented. It is at least as easy to assume that they evolved segmented bodies independently as it is to assume that the ancestral protostome or bilaterian was segmented and that segmentation disappeared in many descendant phyla. The current view is that annelids are grouped with molluscs, brachiopods and several other phyla that have lophophores (fan-like feeding structures) and/or trochophore larvae as members of Lophotrochozoa. Bryozoa may be the most basal phylum (the one that first became distinctive) within the Lophotrochozoa, and the relationships between the other members are not yet known. Arthropods are now regarded as members of the Ecdysozoa ("animals that molt"), along with some phyla that are unsegmented.

The "Lophotrochozoa" hypothesis is also supported by the fact that many phyla within this group, including annelids, molluscs, nemerteans and flatworms, follow a similar pattern in the fertilized egg's development. When their cells divide after the 4-cell stage, descendants of these 4 cells form a spiral pattern. In these phyla the "fates" of the embryo's cells, in other words the roles their descendants will play in the adult animal, are the same and can be predicted from a very early stage. Hence this development pattern is often described as "spiral determinate cleavage".

Mollusca

Cornu aspersum (formerly *Helix aspersa*) – a common land snail

The mollusc (or mollusk) composes the large phylum Mollusca of invertebrate animals. Around 85,000 extant species of molluscs are recognized. Molluscs are the largest marine phylum, comprising about 23% of all the named marine organisms. Numerous molluscs also live in freshwater and terrestrial habitats. They are highly diverse, not just in size and in anatomical structure, but also in behaviour and in habitat. The phylum is typically divided into 9 or 10 taxonomic classes, of which two are entirely extinct. Cephalopod molluscs, such as squid, cuttlefish and octopus, are among the most neurologically advanced of all invertebrates—and either the giant squid or the colossal squid is the largest known invertebrate species. The gastropods (snails and slugs) are by far the most numerous molluscs in terms of classified species, and account for 80% of the total.

The three most universal features defining modern molluscs are a mantle with a significant cavity used for breathing and excretion, the presence of a radula (except for bivalves), and the structure of the nervous system. Other than these things, molluscs express great morphological diversity, so many textbooks base their descriptions on a "hypothetical ancestral mollusc". This has a single, "limpet-like" shell on top, which is made of proteins and chitin reinforced with calcium carbonate, and is secreted by a mantle covering the whole upper surface. The underside of the animal consists of a single muscular "foot". Although molluscs are coelomates, the coelom tends to be small. The main body cavity is a hemocoel through which blood circulates; their circulatory systems are mainly open. The "generalized" mollusc's feeding system consists of a rasping "tongue", the radula, and a complex digestive system in which exuded mucus and microscopic, muscle-powered "hairs" called cilia play various important roles. The generalized mollusc has two paired nerve cords, or three in bivalves. The brain, in species that have one, encircles the esophagus. Most molluscs have eyes, and all have sensors to detect chemicals, vibrations, and touch. The simplest type of molluscan reproductive system relies on external fertilization, but more complex variations occur. All produce eggs, from which may emerge trochophore larvae, more complex veliger larvae, or miniature adults.

Good evidence exists for the appearance of gastropods, cephalopods and bivalves in the Cambrian period 541 to 485.4 million years ago. However, the evolutionary history both of molluscs' emergence from the ancestral Lophotrochozoa and of their diversification into the well-known living and fossil forms are still subjects of vigorous debate among scientists.

Molluscs have been and still are an important food source for anatomically modern humans. However there is a risk of food poisoning from toxins which can accumulate in certain molluscs under specific conditions, and because of this, many countries have regulations to reduce this risk. Molluscs have, for centuries, also been the source of important luxury goods, notably pearls, mother of pearl, Tyrian purple dye, and sea silk. Their shells have also been used as money in some preindustrial societies.

Mollusc species can also represent hazards or pests for human activities. The bite of the blue-ringed octopus is often fatal, and that of *Octopus apollyon* causes inflammation that can last for over a month. Stings from a few species of large tropical cone shells can also kill, but their sophisticated, though easily produced, venoms have become important tools in neurological research. Schistosomiasis (also known as bilharzia, bilharziosis or snail fever) is transmitted to humans via water snail hosts, and affects about 200 million people. Snails and slugs can also be serious agricultural pests, and accidental or deliberate introduction of some snail species into new environments has seriously damaged some ecosystems.

Etymology

The words mollusc and mollusk are both derived from the French *mollusque*, which originated from the Latin *molluscus*, from *mollis*, soft. *Molluscus* was itself an adaptation of Aristotle's (*ta malaká*), "the soft things", which he applied to cuttlefish. The scientific study of molluscs is accordingly called malacology.

The name Molluscoida was formerly used to denote a division of the animal kingdom containing the brachiopods, bryozoans, and tunicates, the members of the three groups having been supposed to somewhat resemble the molluscs. As it is now known these groups have no relation to molluscs, and very little to one another, the name Molluscoida has been abandoned.

Definition

The most universal features of the body structure of molluscs are a mantle with a significant cavity used for breathing and excretion, and the organization of the nervous system. Many have a calcareous shell.

Molluscs have developed such a varied range of body structures, it is difficult to find synapomorphies (defining characteristics) to apply to all modern groups. The most general characteristic of molluscs is they are unsegmented and bilaterally symmetrical. The following are present in all modern molluscs:

- The dorsal part of the body wall is a mantle (or pallium) which secretes calcareous spicules, plates or shells. It overlaps the body with enough spare room to form a mantle cavity.

- The anus and genitals open into the mantle cavity.

- There are two pairs of main nerve cords.

Other characteristics that commonly appear in textbooks have significant exceptions:

Supposed universal Molluscan characteristic	Whether characteristic is found in these classes of Molluscs						
	Aplacophora	Polyplacophora	Monoplacophora	Gastropoda	Cephalopoda	Bivalvia	Scaphopoda
Radula, a rasping "tongue" with chitinous teeth	Absent in 20% of Neomeniomorpha	Yes	Yes	Yes	Yes	No	Internal, cannot extend beyond body
Broad, muscular foot	Reduced or absent	Yes	Yes	Yes	Modified into arms	Yes	Small, only at "front" end
Dorsal concentration of internal organs (visceral mass)	Not obvious	Yes	Yes	Yes	Yes	Yes	Yes
Large digestive ceca	No ceca in some Aplacophora	Yes	Yes	Yes	Yes	Yes	No
Large complex metanephridia ("kidneys")	None	Yes	Yes	Yes	Yes	Yes	Small, simple
One or more valves/ shells	Primitive forms, yes; modern forms, no	Yes	Yes	Snails, yes; slugs, mostly yes (internal vestigial)	Octopuses, no; cuttlefish, nautilus, squid, yes	Yes	Yes
Odontophore	Yes	Yes	Yes	Yes	Yes	No	Yes

Diversity

Estimates of accepted described living species of molluscs vary from 50,000 to a maximum of 120,000 species. In 1969 David Nicol estimated the probable total number of living molluscs at 107,000 of which were about 12,000 fresh-water gastropods and 35,000 terrestrial. The Bivalvia would comprise about 14% of the total and the other five classes less than 2% of the living molluscs. In 2009, Chapman estimated the number of described living species at 85,000. Haszprunar in 2001 estimated about 93,000 named species, which include 23% of all named marine organisms. Molluscs are second only to arthropods in numbers of living animal species—far behind the arthropods' 1,113,000 but well ahead of chordates' 52,000. About 200,000 living species in total are estimated, and 70,000 fossil species, although the total number of mollusc species ever to have existed, whether or not preserved, must be many times greater than the number alive today.

Molluscs have more varied forms than any other animal phylum. They include snails, slugs and other gastropods; clams and other bivalves; squids and other cephalopods; and other lesser-known but similarly distinctive subgroups. The majority of species still live in the oceans, from the sea-shores to the abyssal zone, but some form a significant part of the freshwater fauna and the terrestrial ecosystems. Molluscs are extremely diverse in tropical and temperate regions, but can be found at all latitudes. About 80% of all known mollusc species are gastropods. Cephalopoda such as squid, cuttlefish, and octopuses are among the neurologically most advanced of all invertebrates. The giant squid, which until recently had not been observed alive in its adult form, is one of the largest invertebrates, but a recently caught specimen of the colossal squid, 10 m (33 ft) long and weighing 500 kg (1,100 lb), may have overtaken it.

About 80% of all known mollusc species are gastropods
(snails and slugs), including this cowry (a sea snail).

Freshwater and terrestrial molluscs appear exceptionally vulnerable to extinction. Estimates of the numbers of nonmarine molluscs vary widely, partly because many regions have not been thoroughly surveyed. There is also a shortage of specialists who can identify all the animals in any one area to species. However, in 2004 the IUCN Red List of Threatened Species included nearly 2,000 endangered nonmarine molluscs. For comparison, the great majority of mollusc species are marine, but only 41 of these appeared on the 2004 Red List. About 42% of recorded extinctions since the year 1500 are of molluscs, consisting almost entirely of nonmarine species.

Hypothetical Ancestral Mollusc

Because of the great range of anatomical diversity among molluscs, many textbooks start the subject of molluscan anatomy by describing what is called an *archi-mollusc, hypothetical generalized mollusc,* or *hypothetical ancestral mollusc (HAM)* to illustrate the most common features found within the phylum. The depiction is visually rather similar to modern monoplacophorans, and some suggest it may resemble very early molluscs.

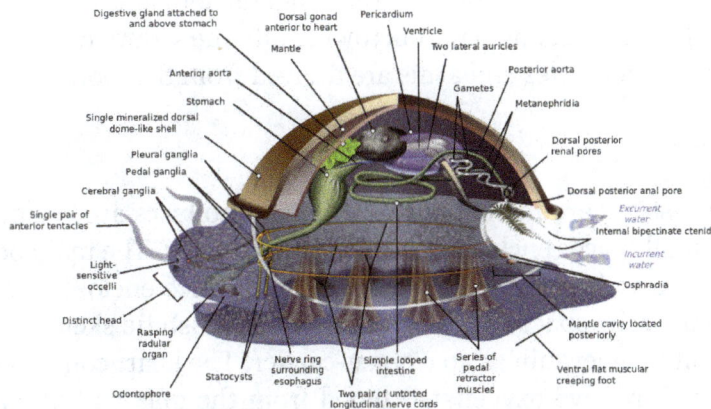

Anatomical diagram of a hypothetical ancestral mollusc

The generalized mollusc is bilaterally symmetrical and has a single, "limpet-like" shell on top. The shell is secreted by a mantle covering the upper surface. The underside consists of a single muscular "foot". The visceral mass, or visceropallium, is the soft, nonmuscular metabolic region of the mollusc. It contains the body organs.

Mantle and Mantle Cavity

The mantle cavity, a fold in the mantle, encloses a significant amount of space. It is lined with epidermis, and is exposed, according to habitat, to sea, fresh water or air. The cavity was at the rear in the earliest molluscs, but its position now varies from group to group. The anus, a pair of osphradia (chemical sensors) in the incoming "lane", the hindmost pair of gills and the exit openings of the nephridia ("kidneys") and gonads (reproductive organs) are in the mantle cavity. The whole soft body of bivalves lies within an enlarged mantle cavity.

Shell

The mantle edge secretes a shell (secondarily absent in a number of taxonomic groups, such as the nudibranchs) that consists of mainly chitin and conchiolin (a protein hardened with calcium carbonate), except the outermost layer in almost all cases is all conchiolin. Molluscs never use phosphate to construct their hard parts, with the questionable exception of *Cobcrephora*. While most mollusc shells are composed mainly of aragonite, those gastropods that lay eggs with a hard shell use calcite (sometimes with traces of aragonite) to construct the eggshells.

The shell consists of three layers: the outer layer (the periostracum) made of organic matter, a middle layer made of columnar calcite, and an inner layer consisting of laminated calcite, often nacreous.

Foot

The underside consists of a muscular foot, which has adapted to different purposes in different classes. The foot carries a pair of statocysts, which act as balance sensors. In gastropods, it secretes mucus as a lubricant to aid movement. In forms having only a top shell, such as limpets, the foot acts as a sucker attaching the animal to a hard surface, and the vertical muscles clamp the shell down over it; in other molluscs, the vertical muscles pull the foot and other exposed soft parts into the shell. In bivalves, the foot is adapted for burrowing into the sediment; in cephalopods it is used for jet propulsion, and the tentacles and arms are derived from the foot.

Circulatory System

Molluscs' circulatory systems are mainly open. Although molluscs are coelomates, their coeloms are reduced to fairly small spaces enclosing the heart and gonads. The main body cavity is a hemocoel through which blood and coelomic fluid circulate and which encloses most of the other internal organs. These hemocoelic spaces act as an efficient hydrostatic skeleton. The blood contains the respiratory pigment hemocyanin as an oxygen-carrier. The heart consists of one or more pairs of atria (auricles), which receive oxygenated blood from the gills and pump it to the ventricle, which pumps it into the aorta (main artery), which is fairly short and opens into the hemocoel.

The atria of the heart also function as part of the excretory system by filtering waste products out of the blood and dumping it into the coelom as urine. A pair of nephridia ("little kidneys") to the rear of and connected to the coelom extracts any re-usable materials from the urine and dumps additional waste products into it, and then ejects it via tubes that discharge into the mantle cavity.

Respiration

Most molluscs have only one pair of gills, or even only one gill. Generally, the gills are rather like feathers in shape, although some species have gills with filaments on only one side. They divide the mantle cavity so water enters near the bottom and exits near the top. Their filaments have three kinds of cilia, one of which drives the water current through the mantle cavity, while the other two help to keep the gills clean. If the osphradia detect noxious chemicals or possibly sediment entering the mantle cavity, the gills' cilia may stop beating until the unwelcome intrusions have ceased. Each gill has an incoming blood vessel connected to the hemocoel and an outgoing one to the heart.

Eating, Digestion and Excretion

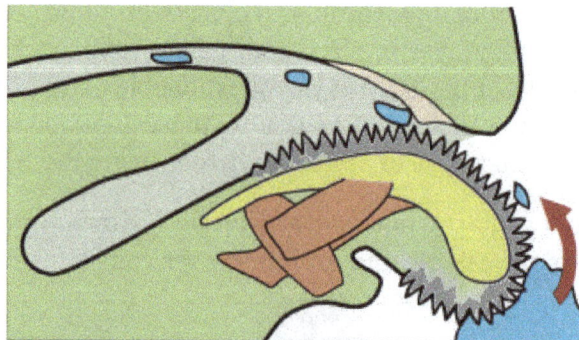

Snail radula at work

| = Food | | = Radula |

= Muscles

= Odontophore "belt"

Members of the mollusc family use intracellular digestion to function. Most molluscs have muscular mouths with radulae, "tongues", bearing many rows of chitinous teeth, which are replaced from the rear as they wear out. The radula primarily functions to scrape bacteria and algae off rocks, and is associated with the odontophore, a cartilaginous supporting organ. The radula is unique to the molluscs and has no equivalent in any other animal.

Molluscs' mouths also contain glands that secrete slimy mucus, to which the food sticks. Beating cilia (tiny "hairs") drive the mucus towards the stomach, so the mucus forms a long string called a "food string".

At the tapered rear end of the stomach and projecting slightly into the hindgut is the prostyle, a backward-pointing cone of feces and mucus, which is rotated by further cilia so it acts as a bobbin, winding the mucus string onto itself. Before the mucus string reaches the prostyle, the acidity of the stomach makes the mucus less sticky and frees particles from it.

The particles are sorted by yet another group of cilia, which send the smaller particles, mainly minerals, to the prostyle so eventually they are excreted, while the larger ones, mainly food, are sent to the stomach's cecum (a pouch with no other exit) to be digested. The sorting process is by no means perfect.

Periodically, circular muscles at the hindgut's entrance pinch off and excrete a piece of the prostyle, preventing the prostyle from growing too large. The anus, in the part of the mantle cavity, is swept by the outgoing "lane" of the current created by the gills. Carnivorous molluscs usually have simpler digestive systems.

As the head has largely disappeared in bivalves, the mouth has been equipped with labial palps (two on each side of the mouth) to collect the detritus from its mucus.

Nervous System

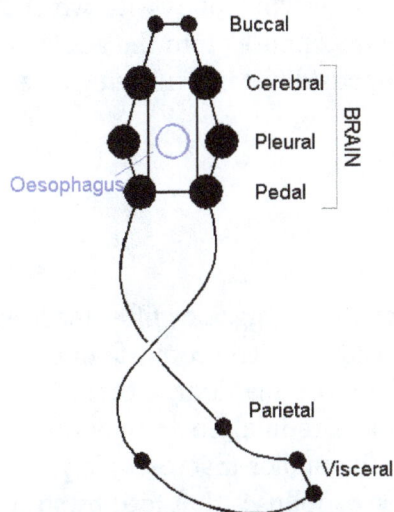

Simplified diagram of the mollusc nervous system

The cephalic molluscs have two pairs of main nerve cords organized around a number of paired ganglia, the visceral cords serving the internal organs and the pedal ones serving the foot. Most pairs of corresponding ganglia on both sides of the body are linked by commissures (relatively large bundles of nerves). The ganglia above the gut are the cerebral, the pleural, and the visceral, which are located above the esophagus (gullet). The pedal ganglia, which control the foot, are below the esophagus and their commissure and connectives to the cerebral and pleural ganglia surround the esophagus in a circumesophageal nerve ring or *nerve collar*.

The acephalic molluscs (i.e., bivalves) also have this ring but it is less obvious and less important. The bivalves have only three pairs of ganglia— cerebral, pedal, and visceral— with the visceral as the largest and most important of the three functioning as the principal center of "thinking". Some such as the scallops have eyes around the edges of their shells which connect to a pair of looped nerves and which provide the ability to distinguish between light and shadow.

Reproduction

The simplest molluscan reproductive system relies on external fertilization, but with more complex variations. All produce eggs, from which may emerge trochophore larvae, more complex veliger larvae, or miniature adults. Two gonads sit next to the coelom, a small cavity that surrounds the heart, into which they shed ova or sperm. The nephridia extract the gametes from the coelom and emit them into the mantle cavity. Molluscs that use such a system remain of one sex all their lives and rely on external fertilization. Some molluscs use internal fertilization and/or are hermaphrodites, functioning as both sexes; both of these methods require more complex reproductive systems.

The most basic molluscan larva is a trochophore, which is planktonic and feeds on floating food particles by using the two bands of cilia around its "equator" to sweep food into the mouth, which uses more cilia to drive them into the stomach, which uses further cilia to expel undigested remains through the anus. New tissue grows in the bands of mesoderm in the interior, so the apical tuft and anus are pushed further apart as the animal grows. The trochophore stage is often succeeded by a veliger stage in which the prototroch, the "equatorial" band of cilia nearest the apical tuft, develops into the velum ("veil"), a pair of cilia-bearing lobes with which the larva swims. Eventually, the larva sinks to the seafloor and metamorphoses into the adult form. While metamorphosis is the usual state in molluscs, the cephalopods differ in exhibiting direct development: the hatchling is a 'miniaturized' form of the adult.

Ecology

Feeding

Most molluscs are herbivorous, grazing on algae or filter feeders. For those grazing, two feeding strategies are predominant. Some feed on microscopic, filamentous algae, often using their radula as a 'rake' to comb up filaments from the sea floor. Others feed on macroscopic 'plants' such as kelp, rasping the plant surface with its radula. To employ this strategy, the plant has to be large enough for the mollusc to 'sit' on, so smaller macroscopic plants are not as often eaten as their larger counterparts. Filter feeders are molluscs that feed by straining suspended matter and food particle from water, typically by passing the water over their gills. Most bivalves are filter feeders.

Cephalopods are primarily predatory, and the radula takes a secondary role to the jaws and tentacles in food acquisition. The monoplacophoran *Neopilina* uses its radula in the usual fashion, but its diet includes protists such as the xenophyophore *Stannophyllum*. Sacoglossan sea-slugs suck the sap from algae, using their one-row radula to pierce the cell walls, whereas dorid nudibranchs and some Vetigastropoda feed on sponges and others feed on hydroids. (An extensive list of molluscs with unusual feeding habits is available in the *"Molluscan diets"*, GRAHAM, A. *(1955) Journal of Molluscan Studies.*)

Classification

Opinions vary about the number of classes of molluscs; for example, the table below shows eight living classes, and two extinct ones. Although they are unlikely to form a clade, some older works combine the Caudofoveata and solenogasters into one class, the Aplacophora. Two of the commonly recognized "classes" are known only from fossils.

Class	Major organisms	Described living species	Distribution
Caudofoveata	worm-like organisms	120	seabed 200–3,000 metres (660–9,840 ft)
Solenogastres	worm-like organisms	200	seabed 200–3,000 metres (660–9,840 ft)
Polyplacophora	chitons	1,000	rocky tidal zone and seabed
Monoplacophora	An ancient lineage of molluscs with cap-like shells	31	seabed 1,800–7,000 metres (5,900–23,000 ft); one species 200 metres (660 ft)
Gastropoda	All the snails and slugs including abalone, limpets, conch, nudibranchs, sea hares, sea butterfly	70,000	marine, freshwater, land
Cephalopoda	squid, octopus, cuttlefish, nautilus, spirula	900	marine
Bivalvia	clams, oysters, scallops, geoducks, mussels	20,000	marine, freshwater
Scaphopoda	tusk shells	500	marine 6–7,000 metres (20–22,966 ft)
Rostroconchia †	fossils; probable ancestors of bivalves	extinct	marine
Helcionelloida †	fossils; snail-like organisms such as *Latouchella*	extinct	marine

Classification into higher taxa for these groups has been and remains problematic. A phylogenetic study suggests the Polyplacophora form a clade with a monophyletic Aplacophora. Additionally, it suggests a sister taxon relationship exists between the Bivalvia and the Gastropoda. Tentaculita may also be in Mollusca.

Fossil Record

Good evidence exists for the appearance of gastropods, cephalopods and bivalves in the Cambrian period 541 to 485.4 million years ago, or even earlier. The first molluscs appear to have been originated earlier. However, the evolutionary history both of the emergence of molluscs from the ancestral group Lophotrochozoa, and of their diversification into the well-known living and fossil forms, is still vigorously debated.

Debate occurs about whether some Ediacaran and Early Cambrian fossils really are molluscs. *Kimberella*, from about 555 million years ago, has been described by some paleontologists as "mollusc-like", but others are unwilling to go further than "probable bilaterian". There is an even sharper debate about whether *Wiwaxia*, from about 505 million years ago, was a mollusc, and much of this centers on whether its feeding apparatus was a type of radula or more similar to that of some polychaete worms. Nicholas Butterfield, who opposes the idea that *Wiwaxia* was a mollusc, has written that earlier microfossils from 515 to 510 million years ago are fragments of a genuinely mollusc-like radula. This appears to contradict the concept that the ancestral molluscan radula was mineralized.

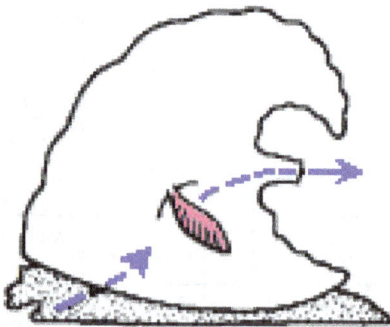

The tiny Helcionellid fossil *Yochelcionella* is thought to be an early mollusc.

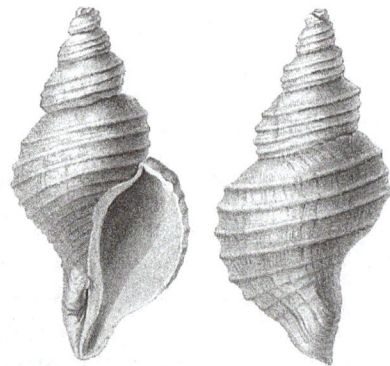

Spirally coiled shells appear in many gastropods.

However, the Helcionellids, which first appear over 540 million years ago in Early Cambrian rocks from Siberia and China, are thought to be early molluscs with rather snail-like shells. Shelled molluscs therefore predate the earliest trilobites. Although most helcionellid fossils are only a few millimeters long, specimens a few centimeters long have also been found, most with more limpet-like shapes. The tiny specimens have been suggested to be juveniles and the larger ones adults.

Some analyses of helcionellids concluded these were the earliest gastropods. However, other scientists are not convinced these Early Cambrian fossils show clear signs of the torsion that identifies modern gastropods twists the internal organs so the anus lies above the head.

Volborthella, some fossils of which predate 530 million years ago, was long thought to be a cephalopod, but discoveries of more detailed fossils showed its shell was not secreted, but built from grains of the mineral silicon dioxide (silica), and it was not divided into a series of compartments by septa as those of fossil shelled cephalopods and the living *Nautilus* are. *Volborthella*'s classification is uncertain. The Late Cambrian fossil *Plectronoceras* is now thought to be the earliest clearly cephalopod fossil, as its shell had septa and a siphuncle, a strand of tissue that *Nautilus* uses to remove water from compartments it has vacated as it grows, and which is also visible in fossil ammonite shells. However, *Plectronoceras* and other early cephalopods crept along the seafloor instead of swimming, as their shells contained a "ballast" of stony deposits on what is thought to be the underside, and had stripes and blotches on what is thought to be the upper surface. All cephalopods with external shells except the nautiloids became extinct by the end of the Cretaceous period 65 million years ago. However, the shell-less Coleoidea (squid, octopus, cuttlefish) are abundant today.

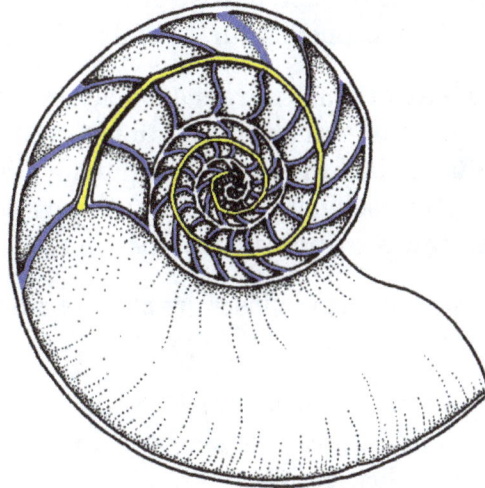

Septa and siphuncle in nautiloid shell

The Early Cambrian fossils *Fordilla* and *Pojetaia* are regarded as bivalves. "Modern-looking" bivalves appeared in the Ordovician period, 488 to 443 million years ago. One bivalve group, the rudists, became major reef-builders in the Cretaceous, but became extinct in the Cretaceous–Paleogene extinction event. Even so, bivalves remain abundant and diverse.

The Hyolitha are a class of extinct animals with a shell and operculum that may be molluscs. Authors who suggest they deserve their own phylum do not comment on the position of this phylum in the tree of life.

Phylogeny

A possible "family tree" of molluscs (2007). Does not include annelid worms as the analysis concentrated on fossilizable "hard" features.

The phylogeny (evolutionary "family tree") of molluscs is a controversial subject. In addition to the debates about whether *Kimberella* and any of the "halwaxiids" were molluscs or closely related to molluscs, debates arise about the relationships between the classes of living molluscs. In fact, some groups traditionally classified as molluscs may have to be redefined as distinct but related.

Molluscs are generally regarded members of the Lophotrochozoa, a group defined by having trochophore larvae and, in the case of living Lophophorata, a feeding structure called a lophophore. The other members of the Lophotrochozoa are the annelid worms and seven marine phyla.

Because the relationships between the members of the family tree are uncertain, it is difficult to identify the features inherited from the last common ancestor of all molluscs. For example, it is uncertain whether the ancestral mollusc was metameric (composed of repeating units)—if it was, that would suggest an origin from an annelid-like worm. Scientists disagree about this: Giribet and colleagues concluded, in 2006, the repetition of gills and of the foot's retractor muscles were later developments, while in 2007, Sigwart concluded the ancestral mollusc was metameric, and it had a foot used for creeping and a "shell" that was mineralized. In one particular branch of the family tree, the shell of conchiferans is thought to have evolved from the spicules (small spines) of aplacophorans; but this is difficult to reconcile with the embryological origins of spicules.

The molluscan shell appears to have originated from a mucus coating, which eventually stiffened into a cuticle. This would have been impermeable and thus forced the development of more sophisticated respiratory apparatus in the form of gills. Eventually, the cuticle would have become mineralized, using the same genetic machinery (engrailed) as most other bilaterian skeletons. The first mollusc shell almost certainly was reinforced with the mineral aragonite.

The evolutionary relationships 'within' the molluscs are also debated, and the diagrams show two widely supported reconstructions:

Morphological analyses tend to recover a conchiferan clade that receives less support from molecular analyses, although these results also lead to unexpected paraphylies, for instance scattering the bivalves throughout all other mollusc groups.

However, an analysis in 2009 using both morphological and molecular phylogenetics comparisons concluded the molluscs are not monophyletic; in particular, Scaphopoda and Bivalvia are both separate, monophyletic lineages unrelated to the remaining molluscan classes; the traditional phylum Mollusca is polyphyletic, and it can only be made monophyletic if scaphopods and bivalves are excluded. A 2010 analysis recovered the traditional conchiferan and aculiferan groups, and showed molluscs were monophyletic, demonstrating that available data for solenogastres was contaminated. Current molecular data are insufficient to constrain the molluscan phylogeny, and since the methods used to determine the confidence in clades are prone to overestimation, it is risky to place too much emphasis even on the areas of which different studies agree. Rather than eliminating unlikely relationships, the latest studies add new permutations of internal molluscan relationships, even bringing the conchiferan hypothesis into question.

Human Interaction

For millennia, molluscs have been a source of food for humans, as well as important luxury goods, notably pearls, mother of pearl, Tyrian purple dye, sea silk, and chemical compounds. Their shells have also been used as a form of currency in some preindustrial societies. A number of species of molluscs can bite or sting humans, and some have become agricultural pests.

Uses by Humans

Molluscs, especially bivalves such as clams and mussels, have been an important food source since at least the advent of anatomically modern humans, and this has often resulted in overfishing. Other commonly eaten molluscs include octopuses and squids, whelks, oysters, and scallops. In 2005, China accounted for 80% of the global mollusc catch, netting almost 11,000,000 tonnes (11,000,000 long tons; 12,000,000 short tons). Within Europe, France remained the industry leader. Some countries regulate importation and handling of molluscs and other seafood, mainly to minimize the poison risk from toxins that can sometimes accumulate in the animals.

Most molluscs with shells can produce pearls, but only the pearls of bivalves and some gastropods, whose shells are lined with nacre, are valuable. The best natural pearls are produced by marine pearl oysters, *Pinctada margaritifera* and *Pinctada mertensi*, which live in the tropical and subtropical waters of the Pacific Ocean. Natural pearls form when a small foreign object gets stuck between the mantle and shell.

Saltwater pearl oyster farm in Seram, Indonesia

The two methods of culturing pearls insert either "seeds" or beads into oysters. The "seed" method uses grains of ground shell from freshwater mussels, and overharvesting for this purpose has endangered several freshwater mussel species in the southeastern United States. The pearl industry is so important in some areas, significant sums of money are spent on monitoring the health of farmed molluscs.

Byzantine Emperor Justinian I clad in Tyrian purple and wearing numerous pearls

Other luxury and high-status products were made from molluscs. Tyrian purple, made from the ink glands of murex shells, "... fetched its weight in silver" in the fourth century BC, according to Theopompus. The discovery of large numbers of *Murex* shells on Crete suggests the Minoans may have pioneered the extraction of "imperial purple" during the Middle Minoan period in the 20th–18th centuries BC, centuries before the Tyrians. Sea silk is a fine, rare, and valuable fabric produced from the long silky threads (byssus) secreted by several bivalve molluscs, particularly *Pinna nobilis*, to attach themselves to the sea bed. Procopius, writing on the Persian wars *circa* 550 CE, "stated that the five hereditary satraps (governors) of Armenia who received their insignia from the Roman Emperor were given chlamys (or cloaks) made from *lana pinna*. Apparently, only the ruling classes were allowed to wear these chlamys."

Mollusc shells, including those of cowries, were used as a kind of money (shell money) in several preindustrial societies. However, these "currencies" generally differed in important ways from the standardized government-backed and -controlled money familiar to industrial societies. Some shell "currencies" were not used for commercial transactions, but mainly as social status displays at important occasions, such as weddings. When used for commercial transactions, they functioned as commodity money, as a tradable commodity whose value differed from place to place, often as a result of difficulties in transport, and which was vulnerable to incurable inflation if more efficient transport or "goldrush" behavior appeared.

Bioindicators

Bivalve molluscs are used as bioindicators to monitor the health of aquatic environments in both fresh water and the marine environments. Their population status or structure, physiology, behaviour or the level of contamination with elements or compounds can indicate the state of contamination status of the ecosystem. They are particularly useful since they are sessile so that they are representative of the environment where they are sampled or placed.

Harmful to Humans

Stings and Bites

The blue-ringed octopus's rings are a warning signal; this octopus is alarmed, and its bite can kill.

Some molluscs sting or bite, but deaths from mollusc venoms total less than 10% of those from jellyfish stings.

All octopuses are venomous, but only a few species pose a significant threat to humans. Blue-ringed octopuses in the genus *Hapalochlaena*, which live around Australia and New Guinea, bite humans only if severely provoked, but their venom kills 25% of human victims. Another tropical species, *Octopus apollyon*, causes severe inflammation that can last for over a month even if treated correctly, and the bite of *Octopus rubescens* can cause necrosis that lasts longer than one month if untreated, and headaches and weakness persisting for up to a week even if treated.

Live cone snails can be dangerous to shell collectors, but are useful to neurology researchers.

All species of cone snails are venomous and can sting painfully when handled, although many species are too small to pose much of a risk to humans, and only a few fatalities have been reliably reported. Their venom is a complex mixture of toxins, some fast-acting and others slower but deadlier. The effects of individual cone-shell toxins on victims' nervous systems are so precise as to be useful tools for research in neurology, and the small size of their molecules makes it easy to synthesize them.

Disease Vectors

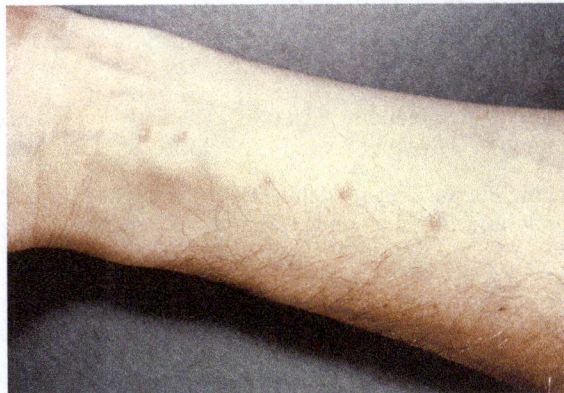

Skin vesicles created by the penetration of *Schistosoma*.

Schistosomiasis (also known as bilharzia, bilharziosis or snail fever), a disease caused by the fluke worm *Schistosoma*, is "second only to malaria as the most devastating parasitic disease in tropical countries. An estimated 200 million people in 74 countries are infected with the disease – 100 million in Africa alone." The parasite has 13 known species, two of which infect humans. The parasite itself is not a mollusc, but all the species have freshwater snails as intermediate hosts.

Pests

Some species of molluscs, particularly certain snails and slugs, can be serious crop pests, and when introduced into new environments, can unbalance local ecosystems. One such pest, the giant African snail *Achatina fulica*, has been introduced to many parts of Asia, as well as to many islands in the Indian Ocean and Pacific Ocean. In the 1990s, this species reached the West Indies. Attempts to control it by introducing the predatory snail *Euglandina rosea* proved disastrous, as the predator ignored *Achatina fulica* and went on to extirpate several native snail species, instead.

References

- Afzelius, BA. "The fine structure of the cilia from ctenophore swimming-plates". The Journal of Biophysical and Biochemical Cytolog. 9: 383–94. PMC 2224992. PMID 13681575. doi:10.1083/jcb.9.2.383

- Hinde, R.T., (1998). "The Cnidaria and Ctenophora". In Anderson, D.T. Invertebrate Zoology. Oxford University Press. pp. 28–57. ISBN 0-19-551368-1

- Horita, T. (March 2000). "An undescribed lobate ctenophore, Lobatolampea tetragona gen. nov. & spec. nov., representing a new family, from Japan". Zoologische Mededelingen. 73 (30): 457–464. Retrieved 2009-01-03

- Tang, F.; Bengtson, S.; Wang, Y.; Wang, X. L.; Yin, C. Y. (20 September 2011). "Eoandromeda and the origin of Ctenophora". Evolution & Development. 13 (5): 408–414. doi:10.1111/j.1525-142X.2011.00499.x

- Tamm, Sidney L. (1973). "Mechanisms of Ciliary Co-ordination in Ctenophores". Journal of Experimental Biology. 59: 231–245

- Wrobel, David; Mills, Claudia (2003) [1998]. Pacific Coast Pelagic Invertebrates: A Guide to the Common Gelatinous Animals. Sea Challengers and Monterey Bay Aquarium. p. 108. ISBN 0-930118-23-5

- Haddock, S.H.D.; Case, J.F. (1995). "Not All Ctenophores Are Bioluminescent: Pleurobrachia" (PDF). Biological Bulletin. 189 (3): 356–362. JSTOR 1542153. doi:10.2307/1542153. Retrieved 2009-02-10

- Haddock, S.H.D. (December 2007). "Comparative feeding behavior of planktonic ctenophores". Integrative and Comparative Biology. 47 (6): 847–853. PMID 21669763. doi:10.1093/icb/icm088

- Tamm, S.L.; Tamm, S. (1985). "Visualization of changes in ciliary tip configuration caused by sliding displacement of microtubules in macrocilia of the ctenophore Beroe". Journal of Cell Science. 79: 161–179. PMID 3914479

- Simon Conway Morris, Alberto M. Simonetta, ed. (1991). The Early Evolution of Metazoa and the Significance of Problematic Taxa. Cambridge University Press. p. 308. ISBN 0-521-11158-7

- Wallberg, A.; Thollesson, M.; Farris, J. S.; Jondelius, U. (2004). "The phylogenetic position of the comb jellies (Ctenophora) and the importance of taxonomic sampling". Cladistics. 20 (6): 558–578. doi:10.1111/j.1096-0031.2004.00041.x

- Arai, M.N. (2005). "Predation on pelagic coelenterates: a review" (PDF). Journal of the Marine Biological Association of the United Kingdom. 85 (3): 523–536. doi:10.1017/S0025315405011458. Retrieved 2009-02-12

- Ehlers, U.; Sopott-Ehlers, B. (June 1995). "Plathelminthes or Platyhelminthes?". Hydrobiologia. 305: 1. ISBN 9789401100458. doi:10.1007/BF00036354

- Nielsen, C.; Scharff, N.; Eibye-Jacobsen, D. (April 1996). "Cladistic analyses of the animal kingdom". Biological Journal of the Linnean Society. 57 (4): 385–410. doi:10.1006/bijl.1996.0023

- Rouse, G.W. (2002). "Annelida (Segmented Worms)". Encyclopedia of Life Sciences. John Wiley & Sons, Ltd. doi:10.1038/npg.els.0001599

- Barker, G.M. (2004). "Terrestrial planarians". Natural Enemies of Terrestrial Molluscs. CABI Publishing. pp. 261–263. ISBN 0-85199-319-2

- Northrop-Clewes, C.A.; Shaw, C. (2000). "Parasites" (PDF). British Medical Bulletin. 56 (1): 193–208. PMID 10885116. doi:10.1258/0007142001902897. Retrieved 2008-12-24

Permissions

Index

www.ingramcontent.com/pod-product-compliance
Lightning Source LLC
Chambersburg PA
CBHW082026190326

41458CB00010B/3285

* 9 7 8 1 6 4 1 1 6 0 1 7 9 *